Mystical Language
in the Later Middl

Medieval History and Culture
Volume 14

STUDIES IN

MEDIEVAL HISTORY AND CULTURE

edited by
Francis G. Gentry
Professor of German
Pennsylvania State University

Routledge
Taylor & Francis Group
LONDON AND NEW YORK

OTHER BOOKS IN THIS SERIES

1. "AND THEN THE END WILL COME"
Early Latin Christian Interpretations of the Opening of the Seven Seals
Douglas W. Lumsden

2. TOPOGRAPHIES OF GENDER IN MIDDLE HIGH GERMAN ARTHURIAN ROMANCE
Alexandra Sterling-Hellenbrand

3. CHRISTIAN, SARACEN AND GENRE IN MEDIEVAL FRENCH LITERATURE
Imagination and Cultural Interaction in the French Middle Ages
Lynn Tarte Ramey

4. WORD OUTWARD
Medieval Perspectives on the Entry into Language
Corey Marvin

5. JUSTICE & THE SOCIAL CONTEXT OF EARLY MIDDLE HIGH GERMAN LITERATURE
Robert G. Sullivan

6. MARRIAGE FICTIONS IN OLD FRENCH SECULAR NARRATIVES, 1170–1250
A Critical Re-evaluation of the Courtly Love Debates
Keith Nickolaus

7. WHERE TROUBADOURS WERE BISHOPS
The Occitania of Folc of Marseille (c. 1150–1231)
Nichole M. Schulman

8. JOHN CASSIAN AND THE READING OF EGYPTIAN MONASTIC CULTURE
Steven D. Driver

9. CHOOSING NOT TO MARRY
Women and Autonomy in the Katherine Group
Julie Hassel

10. FEMININE FIGURAE
Representations of Gender in Religious Texts by Medieval German Women Writers
Rebecca L.R. Garber

11. BODIES OF PAIN
Suffering in the Works of Hartmann von Aue
Scott E. Pincikowski

12 THE LITERAL SENSE AND THE GOSPEL OF JOHN IN LATE MEDIEVAL COMMENTARY AND LITERATURE
Mark Hazard

13. THE REPRODUCTIVE UNCONSCIOUS IN LATE MEDIEVAL AND EARLY MODERN ENGLAND
Jennifer Wynne Hellwarth

Mystical Language of Sensation in the Later Middle Ages

Gordon Rudy

Routledge
Taylor & Francis Group
LONDON AND NEW YORK

First published 2002 by
Routledge

Published 2013 by Routledge
2 Park Square, Milton Park, Abingdon, Oxfordshire OX14 4RN
711 Third Avenue, New York, NY 10017

First issued in paperback 2014

Routledge is an imprint of the Taylor & Francis Group, an informa business

Library of Congress Cataloging-in-Publication Data

Rudy, Gordon.
Mystical language of sensation in the later Middle Ages / by Gordon Rudy.
p. cm. — (Studies in medieval history and culture ; v. 14)
Includes bibliographical references and index.
1. Mystical union—History of doctrines—Middle Ages, 600–1500.
2. Taste—Religious aspects—Christianity—History of doctrines—Middle Ages, 600–1500. 3. Touch—Religious aspects—Christianity—History of doctrines—Middle Ages, 600–1500. 4. Bernard, of Clairvaux, Saint, 1090 or 91–1153. 5. Hadewijch, 13th cent. I. Title. II. Series.
BT767.7 .R83 2002
248.2'2'0902—dc21

2002004216

ISBN 13: 978-0-415-94070-2 (hbk)
ISBN 13: 978-0-415-86702-3 (pbk)

Series Editor Foreword

Far from providing just a musty whiff of yesteryear, research in Medieval Studies enters the new century as fresh and vigorous as never before. Scholars representing all disciplines and generations are consistently producing works of research of the highest caliber, utilizing new approaches and methodologies. Volumes in the Medieval History and Culture series will include studies on individual works and authors of Latin and vernacular literatures, historical personalities and events, theological and philosophical issues, and new critical approaches to medieval literature and culture.

Momentous changes have occurred in Medieval Studies in the past thirty years in teaching as well as in scholarship. Thus the goal of the Medieval History and Culture series is to enhance research in the field by providing an outlet for monographs by scholars in the early stages of their careers on all topics related to the broad scope of Medieval Studies, while at the same time pointing to and highlighting new directions that will shape and define scholarly discourse in the future.

Francis G. Gentry

Contents

Preface

This book is about how some medieval authors wrote about union with God. More specifically, it is about the way they used language that refers to the senses to articulate their ideas about how a person can know God and be one with God. The book is also about the doctrine of the spiritual senses, so-called, which some medieval theologians elaborated to explain how sensory language applies to God and how the bodily people can learn and know nonbodily things. I argue that such explicit concepts of the spiritual senses are not sharply distinct from the ideas implicit in broader usage of sensory language in theological writings.

I focus on language that refers to the senses of touch and taste; I discuss references to sight, hearing, and smell only briefly or in passing. In the past ten years, many articles and some books have appeared that discuss "the senses," and the vast majority discuss only or mainly references to sight and hearing, which they assume can stand in for the other senses. I doubt it. I thought it worthwhile to investigate the history of ideas about the other senses. These ideas are significant in the history of Christian mysticism, because language that refers to the senses bears directly on several ideas that are central to ideas about union with God. Language of taste and touch is especially significant, for reasons that I explain in the first chapter.

Although ideas I examine are theological ideas, this is not a work of theology: that is, I do not interpret anything as a sign of spiritual or divine influence and attempt to figure out what it means, nor do I try to identify human authorities and figure out what they say and mean in order to determine what Christian teachings are, or are best attested, or are best. This book is a history. It is about the writings and ideas of some long-dead people, which interest me as human artifacts. My subject, in other words, is not the subject that interested the medieval people I write about. My topic is not God or mystical union; it is *what people say and think* about God and mystical union—it is people.

I can't imagine that anyone has ever written a book without a lot of help, and I'm delighted to acknowledge the people who have helped me with this one. It began at the University of Chicago Divinity School as a dissertation, and I could not have completed it without the help of my teacher and advisor, Bernard McGinn, who has taught me most of the little I know about medieval Christianity, who guided me through many difficulties as I sieved sources and chapters, and who commented on drafts with unfailing attention and insight. My teachers and readers, Susan Schreiner and Arthur Droge, encouraged me and suggested ways to clarify my arguments and, before I ever set out to write this, taught and inspired me with their historical scholarship. My friends and colleagues Patricia Beckman, Constance Furey, and Edward Howells read and discussed almost every part of the manuscript during three years, and they cheered me through some dark times. Without them I never could have written it. They also made it richer: their own research sparked me to ideas and connections I otherwise would have missed. My parents encouraged me from afar and consistently reassured me that I was not wasting my time. Britt Raphling helped and encouraged me more than she knows, despite everything. Others listened to and discussed parts of my research and their own, especially Joseph Shepley, Amy Thomas, and Donald Duclow; also my colleagues in the Institute for the Advanced Study of Religion at the Divinity School, in the University of Chicago Medieval Studies Workshop, and at the International Congress on Medieval Studies. My teachers, colleagues, and students at the University of Chicago Writing Programs helped me clarify my sentences, paragraphs, and mind. Thank you all. Thanks, last and best, to Calvin Rudy, for stirring me up and cheering me up.

Mystical Language of Sensation in the Later Middle Ages

CHAPTER 1

Sensory Language and Theological Concepts

People use sensory language all the time. It seems inevitable: we are bodies with senses, so we describe and conceptualize the world in bodily and sensory terms. We also, as children, learn languages that are suffused with sensory language. We echo and allude to a vast inheritance of common phrases and ideas that, directly or indirectly, refer to our senses. We can avoid doing so only with great self-consciousness and concentration. It is, therefore, no surprise that Christian writers have used sensory language extensively to discuss God and how we know God. Sensory language is common in the Bible, common in the languages they have spoken and written, and commonly rooted in their bodily existence. However, sensory language has posed a problem for many Christian theologians. How can sensory language, which refers to the acts of perceiving other physical things, possibly apply to a God who is noncorporeal and immaterial? In the history of Christian thought this question is intimately bound up with another: how can bodily humans know and be one with a transcendent God who is wholly beyond the realm of matter and body?

The theological problem posed by sensory language was first addressed explicitly from a Christian perspective by Origen of Alexandria (c. 185-252).[1] He argued that sensory language about God—in particular, the sensory language in the Bible—does not refer to bodily things but necessarily refers to spiritual things. He also made a further claim about the structure of the human person. He argued that sensory language in the Bible refers to "spiritual senses" that belong to the "inner," spiritual person: just as a person has bodily senses that allows them to know material things, so a person has a second set of spiritual senses that allow them to know spiritual things—God, in particular. Origen's idea, known as "the doctrine of the spiritual senses," had a significant influence on late antique and medieval theologians. Because his idea addresses how people can know and be one with God, it is important in the history of Christian mysticism.[2]

Some other Christian theologians have developed similarly explicit and comprehensive solutions to the problem posed by sensory language about God. However, more usually, and especially when discussing mystical union with God, Christian theologians and religious writers have used sensory language without explicitly discussing its appropriateness or significance. That does not mean they have used it at random. Some theologians have used sensory language in orderly patterns that are based on their ideas about physical sensation, their ideas about the structure of the human person that allows us to know and be one with God, and their ideas about God himself. In this book I examine one such pattern of usage as it appears in several significant mystical texts from the twelfth, thirteenth, and fourteenth centuries. The authors of these texts—Bernard of Clairvaux, Hadewijch of Brabant, Bonaventure, Jan Ruusbroec, and a few others—used sensory language of taste and touch to articulate immediate union with God. They did so based on newly prominent ideas about the human person and about the inner nature of God. These theologians tended to blur the distinctions between body and soul or spirit, on the one hand, and between the self and God, on the other. As a result, they implied a concept of "spiritual sensation" significantly different from the concept of spiritual senses taught by other late antique and medieval theologians.

If we want to grasp what is distinctive about the way Bernard, Hadewijch and others used language of taste and touch, and grasp their distinctive concept of the spiritual senses, we must understand something about the alternatives. With reference to the use of sensory language to talk about God, I have already suggested the main alternative: circumscribe it or avoid it. With reference to the concept of the spiritual senses, I will argue that the only significant alternative was the concept first developed by Origen (and later reworked by some scholastic theologians) and the anthropology that supported it. Origen's concept of the spiritual senses is fundamentally exegetical; it is also dualist and intellectualist, by which I mean that he sharply distinguishes matter from spirit and associates spirit with the knowing intellect that "sees." First and foremost, Origen's concept is a doctrine about how to interpret dangerously anthropomorphic language in the Bible, especially in the Song of Songs. Origen is always concerned to counter any bodily or sensual interpretation of the Bible and to refute any notion that God is material and bodily. Therefore, he rejects the material bodiliness implied by sensory language in the Bible, asserting that it refers to spiritual knowledge of God and spiritual contact with him. He supports this exegetical conclusion with the anthropological claim that the sensory language in the Bible refers to the "inner," spiritual person, which has spiritual corollaries to all the organs and members of the "outer" material person. In fact, however, he assimilates all these spiritual corollaries, including the five spiritual senses, to the single "divine sense" by which we can learn and know spiritual things. He identifies the "divine sense" with the knowing intellect that "sees" the divine light. Therefore, Origen rejects all sensory language as too bodily, except language of sight and, to a lesser extent, hearing, which he accepts as references to intellectual knowing and learning—the actions that, he teaches, are among the main tasks for those seeking salvation. By means of the intellect, the person learns from the Logos manifest in the Bible, and as they do their spirit separates from matter and body and ascends to the God who is pure spirit.

Origen's concept of spiritual senses is founded on this anthropological (and cosmic) dualism. He sharply opposes the spiritual senses to corporeal senses because he sharply opposes spirit to material body, and *knowledge* of spiritual things to knowledge of material things. Several later western theologians (some of whom knew Origen's idea) developed similarly dualist concepts of the spiritual senses, if based on somewhat different philosophical ideas. Among the first European theologians to speculate on the concept of the spiritual senses explicitly and at any length were some thirteenth-century scholastics, including Alexander of Hales and Albert the Great. Although they were not very interested in the strongly exegetical character of Origen's concept, they preserved its basic intellectualism and dualism. The scholastics' ideas about anthropology were, of course, significantly different from Origen's Middle Platonic conceptions. They developed the notion of the spiritual senses using Aristotelian concepts of faculty psychology and epistemology and Augustinian ideas about theological anthropology and grace. Nevertheless, they, like Origen, strongly distinguish the corporeal senses from the spiritual senses, which they identify with the knowing intellect. Unlike Origen, they do not identify the intellect with the spirit that, even if it benefits from God's help, is of itself fundamentally and ultimately one with God. Rather, they emphasize (more strongly than Origen did) that we know and are one with God only at his initiative and by his action: both material body and the spirit bound to it are in need of grace. Thus they associate the spiritual senses with graced and "illuminated" intellect. But that does not change the basic dualism of the idea; it just moves the dividing line higher.

However, already in the twelfth century the Cistercian monk Bernard of Clairvaux (1090-1153) had arrived at a less dualist and intellectualist concept of spiritual senses; in the thirteenth century, the Dutch holy woman Hadewijch of Brabant (fl. 1230-1250?) developed a similar idea. Neither of them discuss the concept explicitly, as a formal doctrine, but they imply it in the way they use sensory language to write about God. Their usage departed from established patterns. From Origen's day on, most theologians in the Latin West had agreed that all sensory language is ultimately inappropriate to God. (That is, God the Father. They of course affirmed that Christ was both God and physical man; yet they insisted that his *divinity* was still in no way physical or material.) Most also assumed, as Origen argued, that although all sensory language is ultimately inappropriate to

God, some kinds were more appropriate that others. Language that referred to the senses of sight and hearing seemed appropriate to discuss God and how we know God, first because it is common in the Bible and so seemed authorized, at least if used according to biblical formulas. Medieval theological writers also considered it fitting because they assumed the ancient notion that the senses were ranked from sight and hearing, the most "lofty" and "spiritual," down to taste and touch, the most "lowly" and "bodily."[3] This ranking correlated with their basically dualist anthropology. Although humans are both material body and spirit, our spiritual aspect is divine or potentially divine, while material body is an impure covering, prone to sin and a hindrance to spirit. The ranking also agreed easily with the ancient association between hearing, sight, and the intellect that learns and knows (which Origen emphasizes); and of course, our most spiritual and least bodily senses would be the ones that enable us to know and ascend to God.

Language of touch and taste, on the other hand, was considered by most Christian theologians before the twelfth century to be inappropriate to discuss how we know God. They, even more than the ancients, considered the senses of taste and touch base, having little to do with learning and intellectual knowing and everything to do with the often sinful bodily pleasures of food and sex. Even though language of taste and touch is found in the Bible, most theologians thought it too somatic to discuss God or how we know God and were wary about using language that referred to these "bodily" senses.[4] Two striking examples of theologians who share this wariness, and certainly the most influential in the medieval West, are Augustine (354-430) and Gregory the Great (590-604). As is well known, they both strongly favor language of sight and hearing to discuss how we know and can be one with God. Their preference for these *mediated* senses helps them articulate a theological dualism between spirit and material body that is almost as pronounced as Origen's and almost as intellectualist, although based on a somewhat different ideas about the structure of the human person and about how we come to know and be one with him. I have already mentioned one of their main additions to Origen's idea, which only reinforces this distinction between spirit and body, person and God: they identify spiritual hearing and sight closely with God's action in us—that is, with grace. It is not so much that we hear God, as that he speaks to us; not so much that we see his light, as that he illuminates us with it, making spiritual sight possible. Augustine's and Gregory's language of sight and hearing implies spiritual senses that are less in us, there to be activated, than of God, and added to us as a gift. When they mention the spiritual senses explicitly, which they do in a few places, they assign them to God.

But if most medieval theologians were both wary of sensory language and wary of ascribing any bodiliness to our knowledge of and union with God, some were not, especially after the twelfth century. Bernard of

Clairvaux and Hadewijch use language of taste and touch extensively to articulate an immediate relation between the human person and God precisely because it refers to the embodied, "experienced" self. So do other authors, if somewhat more ambiguously: the Franciscan schoolman St. Bonaventure (1217-1274), the Franciscan Rudolf of Biberach (1270-1330), and the Dutch-speaking contemplative Jan Ruusbroec (1293-1381). All these theologians still usually assume that the most spiritual of the senses are sight and hearing and that taste and touch are the most bodily, and they use language of sight and hearing extensively. But they do not avoid language of taste and touch. Bernard and Hadewijch exploit it to articulate our most intimate and immediate relation to God.[5] They preserve its bodily immediacy because they predicate this language of a literary persona who is an integrated whole, soul and body. Bernard mitigates this somatic immediacy somewhat because he teaches that this integral self still finds God in the act of biblical exegesis. Hadewijch presents a persona who seeks and finds God almost entirely without the mediation of the Bible, and ascribes the fundamental immediacy of bodily contact even to the essential nature of God, which she calls "touch" (*gherienen*).

Behind these five theologians' usage are their theological views about our bond with God and how we realize it—concepts about structure of the human person, of Christ, of the Trinity, and, not least, about the kind of life that allows one to achieve union with God. The crucial pivot is their assumption that our bodies have a central role in our approach to and union with God. All five theologians were clearly influenced by the powerful new emphases in Christian thought and practice during the twelfth and thirteenth centuries—in particular, people's increasing focus on the human Jesus, their hunger for his direct bodily presence in the Eucharist, and their desire to gain holiness and access to God by physically imitating Jesus and his first followers, living the "apostolic life" of suffering service.[6] As André Vauchez has put it, in the years 1180-1220, religious life came to focus "not on monkish contemplation but on the search for conformity to Christ, which would result in the divinization of man *hic et nunc,* in his body and his soul. In this perspective, the ultimate objective of a Christian was no longer the veneration of the King of Heaven in his magnificence but rather the imitation of God who had come to earth in his humanity."[7] The trend first developed in monastic culture (Bernard of Clairvaux is among its exponents) and, in the late twelfth century, was taken up as an ideal by lay people who sought a holy life for themselves and a more perfect Church. The most famous such lay person was Francis of Assisi; another example was Hadewijch. These lay people took up a life of apostolic poverty and service in the world, believing that, by performing such holy work, they could make themselves more virtuous and more intimate with God, as, traditionally, monks had done by means of secluded prayer. They believed that if they imitated earthly work of Christ and his immediate followers in the everyday world, molding their acts and thoughts to the human-and-divine model of Jesus and identifying their own suffering with his suffering on the cross, they would become truly virtuous, perfect themselves, and (some thought) attain exalted knowledge of God and exalted spiritual status. Some of these lay people assumed that, if they could act and work perfectly enough, they would not have to merely hope for intimacy with God after death, in heaven, but could know God and achieve direct contact here and now, in the body.

Supporting these ideas is the belief that, because God took on a human body and was fully human as the man Jesus, he sanctified the human body as a route to knowing God and a means to be one with him. Bernard of Clairvaux believed this, and so did Hadewijch. Because God became a human being, our bodies are in some sense divine and perfectible. Indeed, our bodies are indispensable to achieving union with God. This is a matter of emphasis: neither Bernard nor Hadewijch minimize God's transcendence of the material world (nor do any of the other writers I discuss). But they do vigorously affirm and embrace the bodiliness of God the Son and the immediate union this promises to embodied humans. However, as a result, these writers tend to minimize the distinction between body and spirit in the human person. They assume the person is a "psycho-somatic unity,"[8] equally soul and body. Bernard discusses this idea by talking about the incarnation of Christ, and Hadewijch says more about the shared trinitarian structure of God and the human person; but both assume that we can know and be one with God because of our bodies, not despite them. This seems to be why they are so willing to use highly somatic sensory language of touch and taste to discuss our knowledge of God and union with God. The result is that they distinguish only weakly between the corporeal and spiritual senses. They seem to assume we know and are one with God not by means of a "second set" of senses (as Origen and others taught) but by means of a single sensorium or general activity of sensation, which can be perfected and directed to God but still belongs to the *whole* person, soul and body. For this reason, and to distinguish this idea from more dualist concepts of "spiritual senses," I call this the concept of "spiritual sensation."[9]

This concept and the distinctive use of language that articulates it are my subjects in this book. To show why they are distinctive, I will first outline the major alternative: the "Origenist" concept of the spiritual senses. I will examine Origen's teaching on the subject, then, more briefly, outline Augustine's ideas about spiritual senses and their influence on Gregory the Great, and review the basically similar concepts developed by thirteenth-century scholastic theologians. These scholastics were among the first Latin-speakers to know Origen's writings on the concept of the spiritual senses and to develop it in their own way. Their formulations manifest, to varying degrees, the same fundamental dualism that characterizes Origen's idea and, in some cases, its intellectualism as well.

I will then examine how Bernard of Clairvaux uses language of taste and touch to articulate an immediate relation to God and suggest that this usage assumes a less dualist and nonintellectualist concept of spiritual sensation, rooted in Bernard's incarnational theology and his emphasis on embodied "experience." In the next chapter I will examine the related pattern of sensory language in the writings of Hadewijch and show how she uses it to articulate immediate and fully reciprocal union, partly on the basis of christocentric ideas and partly on the basis of her idea that the human person participates in the inner dynamism of the Trinity. In the final chapter, I will sketch how the mystical writings of Bonaventure, Rudolf of Biberach, and Jan Ruusbroec betray elements of both the "Origenist" concept of the spiritual senses and the alternative concept of "spiritual sensation." Like Bernard and Hadewijch, these writers refer to the senses of taste and touch to articulate immediate union. On the other hand, they emphasize that this relation is not of us, but a gift of God; the spiritual senses are acts of grace, sharply distinct from the corporeal senses and the body.

RHETORIC OF SENSATION

In the following chapters, then, I will examine sensory language in selected medieval mystical texts to argue, in general, that this language articulates theological ideas and, in particular, that it assumes a distinctive concept of spiritual senses. I will say something about formal theological doctrine that is articulated in crisp philosophical categories, especially in the chapter on Origen and the thirteenth-century scholastics. I will say more about the use of sensory language to talk about God and human knowledge of God; more, that is, about sensory *rhetoric*. We will not grasp the full significance of concepts of the spiritual senses if we use explicit theological doctrines to evaluate how theologians use sensory language more generally. Rather the reverse: we understand explicit doctrines of the spiritual senses better, I think, if we interpret them as explicitly rationalized elaborations of the concepts implicit in such usage.

If I begin with patterns of usage, however, I immediately face a problem quite similar to the one that biblical anthropomorphisms posed for Origen: what does sensory language mean, and how? I have already made a couple of key assumptions. First, I have asserted that sensory language is somatic language—that it refers to the human body. This is perhaps open to argument, but the medieval writers I deal with certainly assumed it, especially with regard to the "lower" senses, taste and touch. Second and more arguably, I have assumed that the sensory language that theologians use to talk about relations between people and God is patterned on and refers to the relations between bodies; that is, the relations between the embodied self and other persons and things. Here I follow a number of philosophers, anthropologists, and linguists who have argued that such conceptions about the action of the body in space are fundamental to the way we think and the language we use.[10] Sometimes this reference to bodies is overt; much more often, it is implicit. Sensory language necessarily refers to a sensing body. But how does the referring work? How does sensory language imply a body and relations between bodies?

Hans Jonas' attempted "phenomenology of the senses" offers a useful answer to this question. Jonas tries to analyze how human beings' biologically given physical organism affects the way the human mind organizes the world. He describes how each sense or class of sensations puts us in a certain relation to the world, and each has a cognitive structure that profoundly affects the way we think.[11] Most of Jonas' analysis is not relevant here and it is, of course, debatable, but he makes one key claim that I find persuasive and useful: he distinguishes the senses according to whether they are mediated or unmediated. Thus, hearing and sight are mediated, because they allow us knowledge of things at a distance, by means of a medium other than the body (either disturbed air or light, whether conceived as a substance, as fluctuations in the ether, or whatnot). So, too, the sense of smell requires the mediation of air. On the other hand, the senses of taste and touch are unmediated because they require that the objects known contact the body directly.

This basic distinction is a useful tool to analyze sensory language because it suggests one plausible reason why people refer to one sense rather than another: references to the various senses articulate various relations between themselves and other things, modeled on the relations necessary for the relevant sense to operate. (This is only one of several plausible reasons, because, as I have suggested from the start, we refer to the senses constantly and cannot avoid it.) Most helpfully for the student of Christian thought, Jonas' distinction gives us a way to trace connections between the sensory language of theological texts and key theological concepts about the relation of self and body to other beings. It allows us to grasp how sensory language works as a theological vocabulary.[12]

The distinction between mediated and unmediated senses does not have to be imposed on medieval texts; it is mentioned by some of the writers I examine here. Bernard alludes to precisely this point with reference to touch and union with God; so do Bonaventure and Rudolf of Biberach (quoting Thomas Gallus).[13] That is a happy corroborating gift. The more important point is that the distinction between mediated and unmediated senses agrees well with the way these medieval authors (and the others I will discuss) actually *use* sensory language to talk about God and how we know God. They use language of taste and touch because it implies an immediate relation of bodies, which allows them to articulate the idea that union with God can be immediate. The implication of bodily immediacy is not a problem, because it corresponds with their ideas about theological anthropology. The theologians I wrote about assume we know God not insofar as we are spirit but as whole human persons, as selves who are both soul and body. They assume that we are able to know and be one with him because we are both spirit and matter, as Christ was. Therefore, language of taste and touch gives these theologians a vocabulary to articulate their ideas about both theological anthropology and union with God. They used this "rhetoric of sensation" to explain theological ideas and persuade people to embrace them.[14] (I will explain more about why I consider "rhetoric" the best term in a moment.)

I am not the first to use a "phenomenology of sensation" to analyze the relation between sensory language and theological ideas. Elliot Wolfson has drawn on a phenomenological account of the senses to analyze aspects of how the vision of God is discussed in Jewish mystical texts,[15] and David Chidester has made an extended argument about rhetoric of sight and hearing in Christian religious discourse. As I do, Chidester distinguishes the five senses into two groups, partly on the basis of the ideas of Jonas (and Walter Ong). However, he makes far finer distinctions between the "character" of the senses than I do and develops them into far more complex claims about "religious" discourse. I do not find these persuasive and, therefore, deliberately stick to a very minimal claim about the character of different senses.[16]

SENSORY LANGUAGE, EXPERIENCE, AND THE STUDY OF MYSTICISM

I also keep my attention as much as I can on the language and rhetorical construction of texts, not the "experiences," "perceptions" or "mental states" that they "express." This lets me sidestep some intractable problems with the concept "experience" and to avoid some common assumptions about it that I think have misdirected previous studies of sensory language in theological texts and studies of the spiritual senses, in particular. Almost all scholars who have discussed sensory language and/or concepts of the spiritual senses have assumed that both reveal "experience." Karl Rahner, for example, who wrote first modern study of the doctrine of the spiritual senses, thought we would better understand the explicit doctrine of the spiritual senses if we study sensory metaphors as a general class. He thought that this would allow us to establish which sensory language people use to express which kinds of experience of God (Rahner 1933, 298-99; see also 266). In other words, he treats sensory language, and by extension the doctrine of the spiritual senses, as an artifact that "expresses" prior sensory experience. Others who have studied concepts of spiritual senses have shared his assumption.[17] Hans Urs von Balthasar, another influential theologian who was interested in the concept of the spiritual senses, disagreed with Rahner's account of the spiritual senses on theological grounds but shared the assumption that the doctrine deals with how people experience God (von Balthasar, 2:260-362, esp. 309-26).[18]

Scholars have taken sensory language as evidence of "experience" based on the premise that "experience" is distinct from and prior to "language." Further, they assume they can use written texts to recover and compare not language *about* "religious" or "mystical experience," but such experience *itself*.[19] There is reason to question these assumptions, as philosophers have long known and as some scholars of mysticism began to insist in the 1980s. At this late date, it seems obvious that, however we might define "experience" (an infamously vague term), we cannot neatly segregate it from language. More to the point, even if we grant that "experience" is distinct from and prior to language, it is not clear how we can use written language as evidence for it.

Little of this is news, but the history of the concept "experience" has a rhetorical aspect that is too often overlooked. How do we decide whether a text records experience or appeals to the authority of experience in order to get our attention and persuade us to assent? The answer is not obvious. After all, when we listen and read in everyday life, we usually weigh both options. Consider the example of sensory language. We assume that verbal or written references to sensations, whether specific or general, do transcribe, represent, or somehow record real human sensations—either what happened to the speaker or writer and what they sensed or what happened to and was sensed by someone else. Yet we are also usually aware that such references might not; we know that any speaker or writer might be lying, confused, or mistaken, and/or appealing to experience in order to persuade some audience to accept their authority and ideas.

In this respect, appeals to experience are rhetorically powerful. They have been for a long time. For the student of medieval Christianity, appeals to "experience" are deeply entangled with social and political relations, especially with respect to medieval women. Starting in the twelfth century, at the latest, both clerics and lay people increasingly often used the concept "experience" as a lever to maintain or gain the authority to teach and preach. During this period, the church hierarchy increasingly insisted that only priests were authorized to teach and preach doctrine. Lay people could only encourage others to be pious and direct them to priests. However, increasing numbers of lay people *did* claim the authority to teach and preach, and they frequently supported their claims with discourse that recounted or appeared to be based on "experience" of God. In effect, they claimed to have authority direct from God, bypassing the mediation of church practices and officials (or appearing to bypass it, which was bad enough). Clerics, for their part, permitted some lay people to talk and write about their "experience" of God, because they did believe that from time to time God interacts directly with individuals through the Holy Spirit.

Clerics wanted to hear "what happened" to those blessed by God and proclaim it as evidence of God's grace and power in the world.

However, clerics also bent claims about experience to their own ends. Those ends were only occasionally harmonious and innocent of political motives. Medieval clerics were not a uniform bloc of partisans obedient to "the church": they were a contentious collection of theological and political factions. In general, clerics wanted to control or harness any claim about "experience" of God because it threatened their claim that, for all but a very few saints, the only access to God and salvation was through the ordained officials of the institutional church and the sacramental rituals performed by those officials. Clerics usually insisted that they alone could judge who might have more-privileged private access to God. Many clerics vigorously monitored people's claims to "have the Spirit" or experience God. Some of these claims they quashed, some they judged valid and interpreted in accordance with doctrine—but often against the active resistance of the claimant and of other clerical factions who disagreed with their interpretation or resented the influence gained by a successful claim. Modern scholars eager to recover records of "experience of God" can easily minimize or overlook this struggle and unwittingly support one or the other side.

We should handle appeals to experience with special care when we read about the religious practices and theological ideas of medieval women. From the ancient Mediterranean world to the present, many people have assumed that women are more "experienced" than "rational." With respect to Christian thought and practice, discourse about "experience" has often supported the idea that women are passive vessels who can receive the influence of God on behalf of others, but who may not teach doctrine and *cannot* because they are somehow empty of complex theological ideas.[20] Therefore, not surprisingly, medieval women who wrote about God (and the clerics who wrote for them or about them) very often tried to establish authority for themselves by building their writings around accounts of experience. Even more consistently, then and now, hearers and readers have interpreted such writings as records of experience.[21] Medieval people had some compelling reasons to do so. Medieval men generally assumed that women could not speculate or argue about God in a rational, abstract, formal manner. Even more basically, women were rightly passive, even if not silent, especially in the public realm of theological discourse. Probably many women assumed the same. So a woman's discourse about God was much more likely to gain a hearing and be acceptable if it was presented not as doctrine and intellectual work, but as action and experience, or at least strongly based on experience. This rhetorical technique was not necessarily a self-conscious maneuver. It seems likely that many women accepted and believed that doctrine was not the province of women, and so both spoke about and understood their theological ideas as "experience."

We now, however, still too often overlook these assumptions, both in medieval people and in ourselves. If we accept that writings by and about medieval holy women transcribe these women's "experience" of God, we tend to interpret their writings more as evidence of God than as evidence of a person's mental work. Now we might wish to do so if, like many medieval people, we want to present "holy women" as saintly billboards for God's action in the world. However, that ancient Christian reflex has, all too often, minimized or denied women's intellectual contributions and kept women from the ranks of "thinkers" and, in particular, theologians—or at least "systematic," "real" theologians. It assumes that women are and should be passive, emotional, and nonintellectual and, therefore, sets their words and thoughts apart from and below the penetrating words and thoughts of active, logically rigorous men theologians. It tends to deny the obvious: when women speak and write about God, they do not merely transcribe what happens to them, they do complex intellectual work. Like men, they think about, compare and analyze ideas and concepts. Women speak or write them not to "express" themselves, gushing forth like overheated boilers, but, like men, to achieve something in the world: to solve problems for themselves and others; to teach; to influence; to persuade.

"Experience," then, is a slippery concept, and trying to analyze medieval "experience of God" seems to me to raise enormous problems and solve none of any significance. I think the best object of analysis for the historian of Christianity is, instead, the language and rhetorical construction of texts (which includes claims about experience). We should at least begin there. This starting point is especially important for the study of "mystical" texts. (For my purposes here, I use the phrase "mystical texts" to mean works that discuss how a person seeks and achieves intimate contact, congress, and union with God.) Modern scholars have generally assumed that "mysticism" and "experience" are nearly synonymous. Until recently, they have ignored the fact that this idea became popular only a few hundred years ago. Modern scholars, whether they were interested more in theological content and metaphysics or more in philosophical and psychological analyses, have been most concerned to interpret medieval "mystical" texts in line with their concept of what "mystical experience" is or ought to be, and they have fought endlessly about the "nature" of such experience. This line of argument has been and remains largely a battle between competing theologies and, in some contexts, a defense of theism. Only recently have some scholars begun to concentrate on the more fruitful historical idea that "mystical experience" is not a pure, timeless category that we grope to understand, but an idea with a past.[22]

If, then, we examine mystical texts not as descriptive records of "mystical experience" but as intellectual work, we shift our attention away from weak evidence—the experiences that might have influenced and might be represented in those writings and that are, at best, extremely difficult to compare with other experiences—and toward stronger evidence: what the authors of mystical texts wrote, and how they wrote it. That is my approach in this study. I interpret language of taste and touch in mystical texts as theological rhetoric that articulates concepts. That is, I analyze how it works, both with and sometimes against the assumptions and overall argument of the text, in order to understand both what and *how* the author(s) presents our relation to God in union. I assume the author wrote in order to teach and influence other people who were interested in these subjects. My goal is not to persuade anyone about "what it's like" to "experience God."[23] I want to understand and explain the ideas and thoughts of people in the past who wrote about God.[24]

There is an outstanding example of a study that analyzes the rhetoric of mystical texts in this way. Michael Sells has recently dissected the apophatic "language of unsaying" to show how negative dialectic unsays itself both logically and theologically and how it establishes a relation between God and the person (or literary persona) who speaks, writes, or reads it.[25] Sensory language invites a similar analysis for at least two reasons. First, as I have argued, sensory language is a significant part of what we might call "mystical rhetoric," because it bears on theological issues that are fundamental to mystical texts: anthropology, cosmology, knowledge, and, ultimately, salvation. Second, sensory language has been considered especially important in and especially characteristic of mystical texts. Modern readers have very often assumed that sensory language is direct evidence that describes or transcribes the "mystical experience" they are eager to recover from mystical texts. Accordingly, they have rarely analyzed how it functions as a theological vocabulary or rhetoric of sensation that articulates theological concepts.[26] Modern scholars have, of course, devoted a fair amount of attention to language of hearing and sight. That choice is perhaps justified, since language of sight and light is common in Christian "mystical" texts and is also often found in close conjunction with the negative dialectic that so many people find so compelling. Some scholars have touched on other kinds of sensory language, especially scholars of medieval religious piety and medieval mysticism, who have in recent years increasingly sought to carefully analyze the language of medieval religious texts, including sensory language (e.g., Katz; McGinn 1991, xi-xx, 265-343; and Sharf.).[27] However, when dealing with sensory language, most of these scholars have usually simply assimilated it to explicit doctrines of the spiritual senses and not considered how different patterns of usage agree or disagree with those doctrines. In addition, scholars' interest in language of sight and hearing seems to have blinded them to the way language of taste and touch refers to the fact of body and deafened them to the way theologians exploit this reference to articulate how a person can know and make contact with God—the same basic questions addressed by explicit doctrines of the spiritual senses. I think these questions are the root of the matter.

To help clarify my points about sensory language and experience, let me end with a few words about the perennially popular topic of "synaesthesia." Synaesthetic passages—where an author appeals to two or more senses in a single phrase, clause, or sentence, such that language referring to one sense applies to another—are fairly common in "mystical" texts and are often considered characteristic of them. (We shall find them in the writings of Bernard, Hadewijch, Bonaventure and Ruusbroec.) As I have argued with reference to sensory language generally, I think it is simplistic and misleading to assume such passages express or transcribe some prior "synaesthetic experience" independent of language; that is, some kind of melding or confusion of sensory information in the brain (e.g., sounds that cause a person to perceive a taste, sights that cause them to perceive sounds or tactile sensations—although I am not even sure what those phrases mean). Any discussion of such "synaesthetic experience" raises all the problems I have just discussed with respect to "experience of God," and more. It is more productive, I think, to ignore what we might call "subjective synaesthesia" and pay attention to "objective synaesthesia" or "rhetorical synaesthesia," which I would define as the rhetorical strategy or technique of mixing or "breaking" sensory tropes or metaphors.[28] What is mixed or "broken" are the distinctions between the five senses (or perhaps "classes" of sensation) that people seem to regard as "given" or "natural" (and are, therefore, perhaps inherent to being a sensing body).[29] Understood as a rhetorical technique, we can fruitfully compare synaesthetic rhetoric to negative dialectic or apophatic language. If we can say that broken metaphors are built on contradiction and paradox, it is perhaps not too much to say that like apophatic language, synaesthetic language "unsays" itself. Apophatic language unsays itself in several ways, all built on syntactic and logical *aporia.* As is well known, Christian theologians have exploited apophatic language to discuss the incomprehensibility of the Trinity and the unfathomable otherness of God. I think we can understand synaesthetic language in Christian theological texts in a similar way. By juxtaposing references to one sense with references to others that seem to us distinct, synaesthetic language negates or "unsays" itself.[30] Theologians exploit this to articulate ideas about God and our relation to God.

Again, this approach keeps our attention on the words and ideas in the texts at hand instead of the amorphous concept "experience." It is my approach in this book. I try to show how a few medieval theologians use language of taste and touch to talk about how a person can know God and be one with God. I show that this usage dovetails with these theologians' ideas about Christ, God, and the human capacity to know and be one with God and that it allows them, more broadly, to address the question of how spirit is related to body. I identify patterns of language and trends of thought and suggest how they were influenced by (and themselves influenced) well-known changes in religious practice and theological ideas, especially in the twelfth century and after. That is, I suggest historical change: change in the patterns of sensory language and change in a related idea, the concept that we have spiritual senses. But it remains a suggestion, because I do not pretend to pin down clear causes and later effects (always difficult in a history of ideas), and I examine only a small selection of texts. The writers and texts I have chosen to discuss were written in widely different times, places, and situations, and perhaps these leaps are inexcusable. However, I have not tried to write a comprehensive history or map a chain of influence. Tracing influences is often less informative than it looks, because the really interesting question is not the origin of an idea but why people repeat and develop or recreate an idea in their particular time, place, and situation. I offer only a modest essay in the history of ideas, a kind of morphology balanced around a recurring problem or question. From another perspective, I illuminate one aspect of "mystical language" to help us better understand what some theologians have written, how they wrote it, and a bit of why. I hope this will contribute to the broader study of theological discourse, or at least its mystical corners.

CHAPTER 2

Other Senses

ORIGEN: SPIRIT IN EXEGESIS AND ANTHROPOLOGY

Origen discusses the concept of the spiritual senses frequently in his writings, both early and late.[1] It appears he did so originally to solve the exegetical problem posed by biblical anthropomorphisms. The Bible frequently implies that God is in some way corporeal or material; when we read, for instance, that he worked with his "hands," or that people "saw" God or "heard" his voice.[2] The Song of Songs, in particular, is very sensual and carnal: how is it to be interpreted as referring to God? In common with other Middle Platonic thinkers, Origen assumes that spirit and matter are distinct and opposed and that God is immaterial and incorporeal spirit. How can sensory language refer to him? According to most modern accounts, Origen claims that it refers to the organs and actions of the "inner" person. Just as the "outer," material person has five bodily senses by which we know material things, so the "inner," spiritual person has a set of five spiritual senses by which we can know immaterial, spiritual things. Their primary spiritual object is the Logos, present in the spiritual meaning of the Bible to teach us and make us more thoroughly "spirit." Thus Origen binds interpretation to the structure of the human person, which is itself predicated on the more basic cosmological distinction he observes between spirit and matter.

However, John Dillon has pointed out that it is only in his later works that Origen bends the concept of the spiritual senses to the soteriological distinction between the perfectible "inner" person and the imperfectible "outer" one (Dillon, 443-44). In his earliest and most detailed discussion, in *On First Principles* (c. 230), Origen lays out a more strictly exegetical concept that humans have *a* spiritual sense. Most significantly, Origen associates this single spiritual sense very closely with the operations of the intellect (*mens* or *intellectus; nous; Principles* 1.1.7-9; 1.2.7-8; 2.4.3-4; 4.4.10). Origen claims that "many expressions [in the Bible] which appear to refer to activities of our sense-organs really refer to operations of the mind or soul" as it apprehends the Logos manifest in the Bible (Dillon, 447; see also Harl 1975). More precisely, these expressions refer to the "sight" of the mind, which Origen interprets as the mind's mode or activity of knowing. The *mens* can know its proper spiritual object because both alike are spiritual and not material.

In later works, most prominently in *Commentary on the Song of Songs* and *Against Celsus,* Origen seems more concerned with the details of how this knowing allows the person to become one with God; more profoundly, he is concerned with the precise relationship between matter and spirit in the human person.[3] He recasts his concept of spiritual senses to address these anthropological and soteriological questions. He tries to explain how and where the spiritual senses fit within the human person such that they span matter and spirit and allow us to be transformed. Therefore, he tries to locate the spiritual senses not in the purely spiritual part of the person, but in the more complex coils of the soul (or perhaps between the soul and the spirit), where they can bridge immaterial spirit and material body, "reaching across" from the one to perfect the other.

Interpretations that present this later conception as definitive effectively shade out the strong dualism of his earlier account, or, at best, suggest that he ditched it.[4] I think Origen maintains his earlier concept intact, and the fundamental cosmological and anthropological assumptions that go with it. His later discussions are not especially coherent because, as he did in *On First Principles,* he consistently assumes we can learn and know about immaterial spiritual things (*intellectualia*) only because part of the human person, the intellect, is also immaterial spirit. Spirit can know spirit; matter can not. This leads him to assimilate the five spiritual senses to one, sight, and assimilate it so entirely to the action of spiritual part of the human person, the intellect, that the bodily sensoriness of the spiritual sense(s) vanishes. He still assumes that the *sensus divinus* is no way material or bodily. It is accordingly hard to grasp how it mediates between or "reaches across" from spirit to matter.

There is evidence that this was a persistent problem for Origen, because he has similar difficulties reconciling spirit and matter in his teachings on resurrection and the Incarnation. His teaching on these points is disputed, but it underlines his basic problem: he has trouble maintaining the spiritual sense(s) as a bridge or "medium" between matter and spirit because his basic idea is that the spiritual sense(s) is *spiritual* and not bodily. It must be, to allow him to fix a spiritual meaning for the troublingly corporeal and carnal language of the Bible and, by extension, to explain away any sensory language that might be applied to spiritual things. Admittedly, the problem of reconciling matter and spirit seems inherent to Platonic metaphysics (and, for that matter, other ancient Mediterranean cosmologies); but it has also been a persistent issue for Christian thinkers. Origen was only the first to bind it explicitly to the problem of sensory knowledge, the structure of the human person, and the proper interpretation of language. Indeed, despite their somewhat different philosophical inheritance, the thirteenth-century scholastics who next developed the idea explicitly also arrived at similarly dualist conceptions, even as others in their day were coming up with more a integral version.

Exegesis, Sense, and Intellect

In *On First Principles,* Origen develops his concept of the spiritual senses against the notion that God is a body (*Principles* 1.1.7-9; 2.4.3-4; 4.4.10). He does so using both "philosophical" and biblical language,[5] because he argues both against the opinions of traditional philosophical schools, such as Epicureans and Stoics, "who maintain that first principles are material. . . . even the Logos of God," and against the teachings of some Christians, especially Marcion and his followers (*Celsus* 6.71). Marcionites rejected the Hebrew scriptures, partly because they are filled with sensory, materialist, and anthropomorphic language for God. They concluded that these books could not possibly refer to the spiritual and invisible father of Jesus, but must refer to another, perceptible God. In order to attack this conclusion and support his belief that the Hebrew and the Christian writings refer to a single immaterial and spiritual God, Origen reinterprets the evidence. He argues that the Bible's corporeal and materialist language for God cannot be taken literally. After all, he notes, not only Moses speaks of a visible God,

> for Christ says, 'He who hath seen the Son hath seen the Father also' [Jn 14:9] . . . . [which] we interpret . . . as referring not to sight but to understanding. For he who has understood the Son has understood the Father also. It is in this manner that we must suppose Moses to have seen God, not by looking at him with the eyes of the flesh, but by understanding him with the vision of the heart and the sense of the mind [*visu cordis ac sensu mentis intelligens*], and even this in part only. For it is well-known that he, that is, the one who gave the oracles to Moses, says, 'Thou shalt not see my face, but my back' [Ex 33:23]. Certainly these statements must be understood by the aid of that mystery which is appropriate to the understanding of divine sayings, and those old wives' fables, which ignorant people invent on the subject of the front and back parts of God, must be utterly rejected and despised. (*Principles* 2.4.3; translation Butterworth, slightly altered)

> [. . . uideri deus pater Christi dicitur, quoniam *qui uidit,* inquit, *filium, uidit et patrem* {Jn 14:9}. Quod utique uos quidem uehementer angustat, a nobis uero rectius non pro uidendo sed pro intelligendo sentitur. Qui enim intellexerit filium ille intellexit et patrem. Hoc ergo modo etiam Moyses deum uidisse putandus est, non oculis eum carnalibus intuens, sed uisu cordis ac sensu mentis intelligens, et hoc ex parte ali-

> qua. Manifestum est enim quia *faciem,* inquit, *meam non uidebis* (is scilicet, qui Moysi responda praebebat), *sed posteriora mea.* Quae utique cum eo sunt intellegenda sacramento, quo intellegi conuenit dicta diuina, abiectis profecto illis et spretis anilibus fabulis, quae de anterioribus dei ab imperitis, posterioribusque finguntur.]

In the sacred writings, therefore, "seeing" (*videre*) does not have a materialist, corporeal meaning, but signifies "understanding" (*intelligere*). Here Origen operates according to his usual principles of biblical exegesis. Scholars still argue about what exactly these are, but we can at least say that he distinguishes between an obvious and a nonobvious meaning.[6] The obvious one is the commonplace meanings of the words and grammar in a passage, which he usually calls a "bodily" or "historical" meaning "according to the letter." The nonobvious meaning he calls various things: most basically, the "spiritual" meaning. It is true that Origen says various things about the kinds or levels of meaning. He mentions three levels that correspond to body, soul, and spirit (e.g., *Principles* 4.2.4), and sometimes three levels within the *spiritual* meaning: moral, symbolic, and anagogic (e.g., *Principles* 4.2.5-6). Some scholars have treated this as a firm structure or hierarchy of equally weighted parts.[7] But it is hard not to conclude that for Origen the basic distinction is between the literal or "bodily" meaning and the spiritual meaning in general (or collectively), because he generally fails to maintain a clear distinction between the various "spiritual" meanings. In *On First Principles,* for instance, although he mentions three meanings, he often stresses only that the "bare letter" must be distinguished from its "spiritual sense" and mainly emphasizes how the "bodily" meaning differs from the "allegorical," "mystical," or "spiritual" one (*Principles* 4.2.2, 4.2.4-9).[8]

The nonobvious "spiritual" meaning that is "within" or "beyond" the obvious "bodily" meaning is not secondary, but primary: "all [scripture] has a spiritual meaning, but not all a bodily meaning" (*Principles* 4.3.5). Many biblical passages did not or could not happen, do not clearly refer to Jesus' saving message, are inappropriate to Origen's conception of God, or clash with our perceptions of the physical world (*Principles* 4.2.5; cf. *Celsus* 7.35). Materialist and corporeal language is the extreme instance. It clashes so strongly with the immateriality of spiritual things that when it is applied to God and our knowledge of God, that it can have "no bodily sense," but only a spiritual one.

Origen supports this conclusion with his idea that "the universal content of Scripture is the teaching of Christ the Logos" (the phrase is from Torjesen, 111). The teaching of Logos is present in every part of the Bible; and it is one and without contradiction, because the Logos is. The evangelists and the prophets transcribed the teaching of the eternal Logos as they received it from the Holy Spirit (e.g., *Principles* 4.2.7, 1.3.1).[9] Therefore, it is present not only in the basic doctrines of salvation we easily find in the New Testament, but also in more obscure New Testament passages, and in the "spiritual sense" of the Old Testament (*Principles* 1.3.1, 4.2.2-3). Indeed, Origen claims the Logos is especially present in those passages most inappropriate to God and spiritual things. Because their obvious meaning cannot be accepted, they lead the interpreter to God. They force the exegete to find an immaterial and spiritual referent, and so lure the interpreter beyond the evident material realm to the hidden realm of immaterial spirit (*Principles* 4.2.5).[10]

Origen's problem is to explain how this happens. He implies part of his answer in his claim that we see God "by understanding him with the vision of the heart (*visu cordis*) and the sense of the mind (*sensu mentis*)" (*Principles* 2.4.3). What does he mean by this? In *On First Principles,* we find the answer in the more philosophical components of his argument about corporeal and materialist language. He claims that God is mind (*mens*) and, therefore, immaterial and incorporeal, and that human beings can know this because we too are partly mind. Therefore, material and sensory language that is ascribed to God in the Bible must refer to immaterial mind. Moreover, he claims this is clearly the usage of the Bible, where (as he renders it) Solomon specifically says "You will find a divine sense" (Prov 2:5).[11] He thinks this "divine sense" is or belongs to the intellect (*mens*), by which we discover the spiritual or divine sense of the text—although he does not clearly distinguish between the divine sense of the soul and that of the text (I shall return to this issue in due course).

We can see all these key at the start of *On First Principles,* where we find Origen's longest and most detailed treatment of the spiritual senses (1.1.7-9). He begins with his usual exegetical assertion: although we can read that God is a fire (Dt 4:24) and a spirit (Jn 4:24), we cannot conclude he is a body, because "fire" is a metaphor; and although "spirit" is not, in biblical usage "spirit" is opposed to body (*Principles* 1.1.1, 1.1.2-4). But having said that, he elaborates a more philosophical argument that God cannot be a perceptible body because he is "a simple intellectual existence . . . the mind and the fount from which originates all intellectual existence or mind." Therefore, God is not discernible by the senses, and exists "apart from all corporeal intermixture," because "mind . . . does not need magnitude discernible by the senses, nor bodily shape or color . . . suitable to bodies and matter" (*Principles* 1.1.6; translation from Butterworth, slightly altered). [. . . Deus . . . intellectualis natura simplex . . . et mens ac fons, ex quo initium totius intellectualis naturae uel mentis est. Mens uero ut moueatur uel operatur, non indiget loco corporeo neque sensibili magnitudine uel corporali habitu aut colore, neque alio ullo prorsus indiget horum, quae corporis uel materiae propria sunt. . . . oportet totius corporeae admixtionis alienum una sola, ut ita dixerim, deitatis specie constare.]

To substantiate this conclusion and simultaneously attack a materialist view of God, Origen then makes an analogy to the physical senses. Mind cannot be body, he claims, because we know human mind contemplates invisible things and perceives incorporeal things. Although "each of the bodily senses is appropriately connected with a material substance towards which the particular sense is directed," the "sense of mind" (*sensus mentis*), which is superior to the bodily senses and not bodily, has no material object:

> They do wrong even to God himself . . . [who suppose] that he can be understood through a bodily nature, since according to them that which can be understood or perceived through a body is itself a body; and they are unwilling to have it understood that there is a certain intimacy [*propinquitas*] between the mind and God, of whom the mind is an intellectual image, and that by reason of this fact the mind, if it is purified and separated from bodily matter, is able to have some perception of the divine nature. (*Principles* 1.1.7)
>
> [Vnicuique corporeo sensui substantia quaedam sensibilis subiacet proprie, in quam ipse sensus corporalis intenditur. . . . ad ipsum deum refertur inuria, cum putant eum per naturam corpoream posse intellegi, quo scilicet secundum ipsos corpus sit et illud, quod per corpus potest intellegi uel sentiri; et nolunt hoc intellegi, quod propinquitas quaedam sit menti ad deum, cuius ipsa mens intellectualis imago sit, et per hoc possit aliquid de deitatis sentire natura, maxime si expurgatior at segregatior sit a materia corporali.]

That is, mind is not body because it is like its object (*Principles* 1.1.7).[12] It knows by way of its near relationship or affinity (*propinquitas*), a term Origen does not explain but that seems to refer to the likeness it shares with its original source. Like all Platonists, he assumes we know by means of "likeness." The physical senses perceive their appropriate objects because they are "like" and participate in them (*Principles* 1.1.7).[13] So too does the mind. We can know God by means of the mind because the mind, like the sense organs, knows what is like it. This analogy could imply that the mind's object is a substance or a body. But Origen asserts the mind is a "faculty of an intellectual nature" and, therefore, incorporeal and immaterial. So also it has a like object of intellectual and *spiritual* nature: God.[14]

Just as the physical senses are of but distinct from the body, so it seems the *sensus divinus* should be of but distinct from the intellect of which it is a "power" (*virtus*): "eyes of a pure heart" or "mind" (Rahner 1932, 124, 125; Harl 1975, 25; 26, n. 12). But Origen leaves this distinction vague. He does not explain how *sensus divinus* is different from or more limited than intellect, except when he links it to the state or condition of the graced or "illumined" *mens* (Harl 1975, 30-33, esp. n. 21). I will discuss this important point in a moment. Here I only want to observe that Origen does not distinguish how *sensus divinus* and *mens* differ, he emphasizes how the two are *alike*. A person can know God because each person participates in God.[15] Even though, because of its "negligence" (*negligentia*), intellect to fell from perfect union with him and was "bound" or "clothed" in material body, *mens* still shares in God's incorruptible and eternal intellectual nature (*Principles* 1.4, 1.8, 2.6.3, 2.8, 4.4.9; cf. *Homiliae in Genesim* 1). Human beings retain a "blood-relationship" (*consanguinitatem*) with God because intellect bears an image of the Logos. Therefore, if a person acquires the virtues of God, their intellect can move beyond bodily things and "investigate" or "seek" (*requirere*) intellectual things—in particular, the presence of the Logos in the Bible. Intellect can then attain knowledge of higher spiritual things (*Principles* 4.4.10).

Origen considers these to be the noetic actions of *mens*. Supporting his philosophical explanations with biblical evidence, he interprets biblical references to light, sight and "the heart" as metaphors of knowing and mind. No one can see God, not even Christ, because

> to see and be seen is a property of bodies; [however] to know and to be known is an attribute of intellectual existence. Whatever therefore is proper to bodies must not be believed of either the Father or of the Son, the relations between whom are such as pertain to the nature of deity. . . . what is called 'seeing' and 'being seen' in the case of bodily existence is with the Father and the Son called 'knowing' and 'being known,' through the faculty of knowledge [i.e., the *mens*] and not through our frail sense of sight.[16] (*Principles* 1.1.8)
>
> [. . . naturaliter uideri impossibilis est. . . . Aliud est uidere, aliud cognoscere; uideri et videre corporum res est, cognosci et cognoscere intellectualis naturae est. Quicquid ergo proprium corporum est, hoc nec de patre nec de filio sentiendum est; quod uero ad naturam pertinet deitatis, hoc inter patrem et filium constat. Deinque etiam ipse in euangelio non dixit quia nemo uidit patrem nisi filius neque filium nisi pater, sed ait: *Nemo nouit filium nisi pater, neque patrem quis nouit nisi filius*. Ex quo manifeste indicatur quod quicquid inter corporeas naturas uidere et uideri dicitur, hoc inter patrem et filium cognoscere dicitur et cognosci, per uirtutem scientiae, non per uisibilitatis fragilitatem.]

God is not sensed, but known. He is an "intelligible" accessible only to intellect. All similar references in the Bible to the sensation and the organs of sense are equivalent metaphors. He claims that the Bible itself indicates its sensory language refers to mind, at Prov 2:5, and that biblical usage makes clear that this sense is intellectual (intellect or a property of intellect or mind). Thus, when we read in the Bible that the pure in heart shall see God (Mt 5:8), this does not refer to physical sight:

> For the names of the organs of sense are often applied to the soul, as that we speak of seeing with the eyes of the heart [cf. Eph 1:18], that is, of drawing some intellectual conclusion by the power of intelligence. So too we speak of hearing with the ears when we discern the deeper

> meaning of some statement [cf. Mt 13:9]. So too we speak of the soul as being able to use teeth, when it eats and consumes the bread of life who comes down from heaven [cf. Jn 6:32ff]. In a similar way we speak of it as using all the other bodily organs, which are transferred from their corporeal significance and applied to the powers of the soul; as Solomon says, 'You will find a divine sense' [Prov 2:5]. For he knew that there were in us two kinds of senses, the one being mortal, corruptible and human, and the other immortal and intellectual, which he here calls 'divine'. By this divine sense, therefore, not of the eyes but of a pure heart, that is the mind, God can be seen by those who are worthy. That heart is used for mind, that is for the intellectual faculty, you will certainly find over and over again in all the scriptures, both the New and Old.[17] (*Principles* 1.1.9; translation Butterworth, slightly altered)
>
> [Frequenter namque sensibilium membrorum nomina ad animam referuntur ita, ut oculis cordis videre dicatur, id est uirtute intelligentiae aliquid intellectuale conicere. Sic et audire auribus dicitur, cum sensum intellegentiae profundoris aduertit. Sic et ut eam posse dentibus dicimus, cum mandit et comedit panem uitae, qui de caelo descendit. Similiter et ceteris uti membrorum officiis dicitur, quae ex corporali appellatione translata uirtutibus animae coaptantur, sicut et Salomon dicit: *Sensum diuinum inuenies*. Sciebat namque duo genera esse sensuum in nobis, unum genus immortale et intellectuale, quod nunc divinum nominauit. Hoc ergo sensu diuino non oculorum, sed cordis mundi, quae est mens, deus uideri ab his, qui digni sunt, potest. Cor sane pro mente, id est pro intellectuali uirtute nominari in omnibus scripturis nouis ac veteribus abundanter inuenies.]

The *sensus divinus* is the *mens,* which is the "divine sense" because God is mind. Origen claims the *mens* can "investigate" and "seek" (*requirere*) intellectual things because, in the same way as the bodily senses, the *mens* is like its object. However, its object is no material or bodily thing, but God, who is intelligible, immaterial, and incorporeal. The *sensus divinus* is "spiritual" because it is opposed to material body (*Principles* 1.1.1).

The Senses of the "Inner" Person

In *On First Principles* Origen seems interested in the *sensus divinus* as a constituent or power of mind primarily because it helps him outline the proper "spiritual" meaning of the Bible and so counter materialist and corporeal concepts of God. *Sensus divinus* discovers this spiritual meaning, which manifests the Logos, because it, too, is spirit (*Principles* 2.8.2). Although mind is held in material body, it is not *of* the body and does not need it to know spiritual things; in fact, body is a hindrance (see *Principles* 1.1.7). So the *sensus divinus* seems divorced from the whole person, body and soul, and seems to discover and learn from the Logos independent of the whole person. In his later works, on the other hand, Origen seems more concerned to show how the *sensus divinus* operates within the human person to discover proper spiritual meaning of the Bible and make us spiritual. He elaborates a more explicitly anthropological and soteriological concept of the *sensus divinus* in which it mediates between or bridges matter and spirit, transforming one into the other. However, in the end Origen still assimilates the spiritual senses to intellect and to spirit. *Sensus divinus* renews or transforms us by means of its noetic function, as it allows various manifestations of spirit to recognize and know each other. Matter and body do not have a significant role; *they* are not transformed, they are left behind. This echoes Origen's principle that the sensory and bodily language in the Bible can only have a spiritual meaning, not a bodily one.

Origen presents his concept of the "inner" and "outer" person together with his concept of the spiritual senses in several of his later works, especially *Dialogue with Heraclides, Commentary on the Song of Songs,* and *Against Celsus.* (I treat these together because they are not dated precisely enough to be arranged in sequence.) The fundamental features of this concept are perhaps clearest in the Prologue to the *Commentary on the Song.* Here Origen discusses the spiritual senses to support his argument that the Song of Songs, apparently a profane love poem, actually dramatizes how Christ relates to the church and to the individual soul. In particular, he argues that the sensual and erotic language in the Song refers not to carnal love but to a spiritual, "intellectual" love that draws us to God. To prove this, he claims that the human person, the cosmos and the biblical text share a fundamental distinction between matter and spirit. From Paul he takes the idea that there is an "outer" and "inner" in the human person (see 2 Cor 4:16, Rom 7:22, Col 3:9-10, Eph 3:16). The outer is corrupted while the inner is "renewed from day to day." This moral or soteriological distinction matches a cosmological one. Origen writes that the "inner man" and "outer man" are the same "two men" we read about in Genesis, "the first [made] *in the image and likeness of God,* the second formed *of the slime of the earth* [Gen 1:26, 2:7]." Thus the outer person is visible, physical, and material, while the inner one is invisible, immaterial, and spiritual.

Origen finds this anthropological and cosmological distinction in the way language is used in the Bible. He claims that, in the Bible, different things are given the same names–that is, the Bible makes use of homonyms (*On the Song,* Prol. 2.6). His evidence is the sensory and corporeal language in the Bible, which he insists refers not to the visible but to the invisible body (*On the Song,* Prol. 2.9, 2.11). Therefore, just as references to bodily age (childhood or maturity) also signify the passage from beginner to spiritual perfection,

> so also will you find the names of the members of the body transferred to those of the soul; or rather the faculties and powers of the soul are

> to be called its members. . . . It is perfectly clear that in these passages the names of the members can in no way be applied to the physical body, but must be referred to the parts and powers of the invisible soul. The members have the same names, yes; but the names plainly and without any ambiguity carry meanings proper to the inner, not the outer man. (*On the Song,* Prol. 2.9)
>
> [Igitur sicut haec quae memoravimus aetatum nomina iisdem vocabulis et exteriori homini adscribuntur et interiori, ita invenies etiam membrorum nomina corporalium transferri ad animae membra, seu potius efficientiae haec animae affectusque dicendi sunt. . . . evidenter ostenditur membrorum haec nomina nequaquam corpori visibili aptari posse, sed ad invisibilis animae partes virtutesque debere revocari, quoniam vocabula quidem habent similia, aperte autem et sine ulla ambiguitate non exterioris, sed interioris hominis gerunt.][18]

So too for love. The Song's erotic language only *appears* to refer to a carnal love appropriate to the outer, material person; it actually signifies a spiritual or divine love appropriate to the inner, spiritual person, a love that draws it to God (*On the Song,* Prol. 2.16).

Doesn't Origen just mean these words are metaphors? Yes and no. Yes, because he claims biblical references to the material body have a spiritual meaning, but no, because he claims that this spiritual meaning literally refers to the parts of an inner, spiritual person. In fact, Origen is happy to find a corresponding "spiritual" counterpart for every physical member. He cites biblical texts that mention not just organs of sense—eyes, ears, hands, and tongues—but also feet, wombs, teeth, and arms (*On the Song,* Prol. 2.10), bones, blood, and hair (*Heraclides* 15-24).[19] He lumps the spiritual members together in this way to persuade his readers that the Bible speaks of both real physical people and of an inner person. So it seems that Origen is simply not interested in the particular purpose or significance of these individual spiritual members. The spiritual senses, on the other hand, do have purpose and significance. Biblical references to sense organs and sensation label something that allows the inner person to apprehend the spiritual meaning of the Bible and be transformed by it. In the Prologue to the *Commentary on the Song,* he argues that someone "at the stage of infancy and childhood in their interior life" and "any man who lives only after the flesh. . . . will twist his whole hearing [of love's language, i.e., the Song] away from the inner spiritual man to the outward and carnal . . ." (*On the Song,* Prol. 1.4). [. . . Nec ad capienda quidem verba haec parvula et infantilis interioris hominis aetas admittitur, . . . Si vero aliquis accesserit, qui secundum carnem tantummodo vir est, . . . ab interiore homine ad exteriorem et carnalem virum omnem deflectet auditum. . . .] The bodily meaning of the Song is directed to the outer person, and the spiritual meaning to the inner: "corruptible things are offered to that which is corruptible, while incorruptible things are set before that which cannot be corrupted" (*On the Song,* Prol. 2.12-13) [corruptibili corruptibilia praebentur, incorruptibili vero incorruptibilia proponuntur].[20]

Origen likes to say that the spiritual senses are five discrete "members" belonging to the inner person. Most modern interpreters, starting with Karl Rahner, accept this (Rahner 1932, 119-21). Marguerite Harl, more precisely, has argued that Origen distinguishes between the global faculty of spiritual perception, a spiritual equivalent to the Aristotelian *sensus communis,* and the five organs and actions of that sense with respect to different classes of objects (Harl 1975, 25).[21] She notes that Origen is not much interested in these particular operations with respect to particular objects of sense. He is much more concerned to demonstrate the operations of the single *sensus divinus.* He catalogs particular objects because they are mentioned in the Bible, and he wants to explain away their dangerously corporeal multiplicity.

In fact, Harl shows that Origen assumes he is talking about the operations of a single, global divine sense. He associates the *sensus divinus* with biblical verses that mention the "mouth" (*stoma, os*) and the "heart" (*kardia, cor*) (esp. Ps 118:131 and 2 Cor 6:11). He finds the connection in the way they are said to "open" or "enlarge"—as he interprets it, to receive God (Harl 1975, 36-9).[22] While he could use "mouth" to indicate a particular spiritual sense, instead he uses it to signify the power of the Bridegroom that touches the intellect and illuminates it with a kiss (Sg. 1:1). He uses "heart" to signify the place where the human is able to receive this kiss; this seems to be the intellect. Such kisses are themselves "the moments when [the soul's] intellect perceives the sense [meaning] of doctrine and of inspired texts" (Harl 1975, 19; e.g., *On the Song* 1.1.14-15). So both terms refer to the soul's single, global "divine sense," which Origen identifies with the highest, spiritual part of the human being, the intellect; and its object is the divine, immortal, intelligible things manifest in the spiritual meaning of the Bible (Harl 1975, 32-3, esp. n. 21). (I should note that Harl distinguishes between the intellect and the inspired or illumined intellect that is perfected by grace. I will deal with this more below, but I believe the distinction to be insignificant for my point here, which is simply that Origen does not distinguish "sensing" spiritual things from "knowing" them by the intellect).

Origen certainly likes to itemize biblical references to the organs or actions of five senses; and he calls them, in the plural, "powers of the soul" (*virtutes animae*) and forms or *species* of the single divine sense (*Principles* 1.1.9; *Celsus* 1:48; *On the Song* 1.4.25-26). In his *Commentary on the Song,* he writes that one sense can be active without the others (*On the Song* 1.4.10-11). In this passage his point is that when the Logos seizes or fills (*occupaverat*) all five, nothing else will satisfy, and the (inner) person will be intent only on God, like the blessed in heaven, who delight in all five spiritual senses (*On the Song* 1.4.10, 1.4.14-15). As elsewhere, his evidence is a list of biblical verses that mention the actions, organs, or objects of the spiritual senses.[23] He names a discrete object for each particular sense: "powers" (*virtutes*) of the Logos "that match each single sense according to its nature and capacity (*naturae earum capacitatique*)" (*On the Song* 1.4.10-11).[24] But then he dissolves these distinctions. He does not specify what these different natures and capacities are, he assimilates the particular operations of the "five" senses to the intellect's power of "seeing" and knowing. As he explains later in the text, "the sight by which God is seen is not of the body, but of the mind and spirit. Our Savior Himself marks this distinction in the Gospel by the word that He employs: for He did not say that no man *saw* the Father save the Son, but that no man *knew* the Father save the Son [Mt 11:27]" (*On the Song* 3.13.46; cf. the very similar passages in *Principles* 1.1.8 and 2.4.8). [Visus, quo Deus videtur, non est corporis, sed mentis et spiritus. Quod et ipse Salvator in Evangelio vocabulo proprio distinguens non dixit: Nemo vidit Patrem nisi Filius, sed: *Nemo novit Patrem nisi Filius.*] Likewise he emphasizes the oneness of their object. In Book 2, after another list of "objects" within the Logos, he writes: "All these things are the One, same Word of God, who adapts himself to the sundry tempers of prayer according to these several guises. . . ." (*On the Song* 2.9.14). [Haec autem omnia unum atque idem est Verbum Dei . . . per haec singula affectibus orationis commutantum. . . .]

When Origen names five discrete members or powers by which the human person may "sense" its divine object, it is not because these senses have various different ways of apprehending the Logos, but because the single divine sense is directed to various signs that mark the presence of the (single) Logos in the Bible, all of which it apprehends in the same (one) way: it "sees" and knows. In fact the only discrete objects for the spiritual senses are biblical verses that refer to the actions and organs of the senses, or their physical objects; and as Origen never fails to say, we cannot take these at face value.

Indeed, these passages not only refer to spirit, they *are* spirit, and Origen does not neatly segregate them from the *sensus divinus* in the person. The thing interpreted and the interpreter are fundamentally one; that is the reason interpretation is possible, and that the purpose of interpretation is to recognize and perfect that fundamental likeness. Origen cites Prov 2:5 ("You will find a divine sense") to support *both* his claim that there is or can be a (single) faculty for perceiving spiritual things in the human person, *and* his claim that there is a spiritual meaning in the Bible.[25] He seems not to mind the ambiguity, because he considers the two to be at least isomorphic, and even the same thing—namely, spirit or "spiritual substance."[26] The human person can know the divine sense of the text, which is a manifestation of the spiritual Logos, because part of the person is also spirit. Thus Origen seems to assume the "five" spiritual senses are one, as is their activity in apprehending the spiritual meaning of Scripture. Again, this is because he considers the two isomorphic; but he seems to do so precisely because likeness allows knowledge.

On the other hand, the spiritual sense(s) are not active in people who are "immature" and untrained, which implies difference of some kind. Sight, if not "trained by learning and diligence," will lack the power of discerning good and evil, and will be directed to not to spiritual but to material and carnal things (On the Song 1.9.18).[27] How then do we activate our spiritual divine sense so that we can use it to learn from and be transformed by the Logos? How do we take that first step from body to spirit? Origen urges that the outer senses must be "closed" and "mortified" (*mortificare*) so the inner ones might attain their object (*Celsus* 7.39; *On the Song* 1.4.16). In other words, bodiliness must be set aside. What is left of "sensing"? To "train" the senses of the inner man seems to be, simply, to find the spiritual meaning of the sacred letter: "these other five senses . . . are sought through training and are said to be trained when they investigate the meaning [*intelligentiam*] of things with more accurate and eager keenness" (*On the Song* 1.4.16). [. . . Hos corporeos quinque sensus alios . . . per exercitia quaeruntur et exercitati dicuntur, cum scilicet intelligentiam rerum acumine acriore discutiunt.] In fact, Origen does not really distinguish the act of "sensing spiritually" from its result. We cannot "sense" the spirit (as it is revealed in the Bible) if the spiritual senses are not activated; and the spiritual senses are not active if we cannot "sense" or know the spirit as present in the Bible. "Activating" or "awakening" is not *how* to achieve or activate spiritual knowing and spiritual knowledge, it *is* spiritual knowing and spiritual knowledge.

Origen perhaps manages to step out of this circle when he ascribes this activation or awakening to something like grace. (As we shall see, this tack was later developed by Bonaventure and Ruusbroec.) We see God (that is, know him) because he has healed our power of vision and given us the Spirit of Knowledge and the Spirit of Wisdom (*On the Song* 3.13.45-46). He bestows the kisses that mark the visits of the Word at the appropriate time (*On the Song* 1.1.14-15). Those who lack a spiritual sense do so because "the divine voice is such that it is heard only by those whom the speaker wishes to hear it" (*Celsus* 2.72)."All these things are the One, same Word of God who, adapted to the several aspects of prayer according to these guises, leaves none of the soul's senses devoid of his grace." (*On the Song* 2.9.12; quoted above, pp. 25-6; cf. the similar passage at 1.4.10-11). This could suggest that grace both activates the spiritual senses and is their object. However, if activating the spiritual senses is equivalent to grace, and that activation is itself spiritual knowledge, as I have argued, then again the spiritual senses seem little distinguished among themselves. They seem to be spirit, as intellect is spirit, and as the Logos is spirit, and so one.

A Dualist Anthropology

In his later work Origen ascribes the spiritual senses to the "inner person" and claims they are the senses by which the soul apprehends the spiritual sense or meaning of the Bible. However, he also continues to assume "they" are undifferentiated spirit, and single, and continues to associate them very closely with the *mens*. It seems the *sensus divinus* knows spirit because it is spirit, and is in no sense "between" it and material body. It is therefore hard to grasp how *sensus divinus* helps a person become spiritual.

We can clarify the difficulty if we briefly examine the structure Origen ascribes to the human person, and ask where the spiritual senses fit within it. He wrote no *De anima,* and so leaves a great deal vague, but the major features of his anthropology are clear enough. We have seen that Origen ascribes an "inner" and an "outer" to every person, and that this split corresponds both to the distinction between spirit and matter and to the distinction between the "literal" or "bodily" meaning of Scripture and its "spiritual" meaning. He unambiguously assigns the spiritual senses to the inner person. On the other hand, he also claims (citing Paul, Thess. 5:23) that the person has *three* parts, body, soul, and spirit (*soma, psyche, pneuma*; *corpus, anima, spiritus*), and that these parts correspond to *three* kinds of meaning in or parts of Scripture, the bodily meaning that is its "flesh," its "soul," and "the spiritual law" (*Principles* 4.2.4; cf. 3.4). In this scheme, he assigns the spiritual senses to the soul.

How do we reconcile the two-part with the three-part scheme? It seems easy, because Origen claims "the soul [is] called the 'inner man'" (*Celsus* 7.38). But Henri Crouzel (who neatly outlines Origen's tripartite anthropology), notes that Origen's conception of "soul" (*psyche*) is hard to define precisely because he does not clearly distinguish it from "spirit" (*pneuma*). Rather, Origen treats soul as a two-part composite that mediates between spirit and matter or "flesh." *Psyche* has a "spiritual" aspect that he assimilates to immaterial spirit (*pneuma*), and a lower, "material and corporeal" aspect that he associates with material body (*soma*) (Crouzel 1989, 123-37; on soul, 124-6; likewise Rahner 1932, 123). The lower part is "fleshly" and fallen, but can be perfected or "spiritualized" by the higher part, which itself can be perfected by learning from spirit (e.g., *Principles* 2.8). Confusingly, however, Origen also ascribes "body" (*soma, corpus*) to both material and spiritual entities (Crouzel 1989, 126-28). Not only does the soul seem to have a bodily component, but the resurrected body is "spiritual" (*pneumatikos*). The soul loses its materiality as it "learns" from the spirit, just as the body is to lose its materiality and be spiritualized at the resurrection. I will discuss his argument about resurrection in a moment; for now, I simply want to note that he distinguishes material from spiritual body, just as he distinguishes a material and a spiritual component in the soul.

Origen seems to be trying to work out another bridging intermediary between matter and spirit, outer and inner. It appears he intends the spiritual senses to bridge spirit and matter in the same way. Origen interprets 2 Cor 4:16 ("the inward man is renewed day by day"; cf. Rom 7:22) to mean that the soul is progressively spiritualized as it learns from the Logos (*On the Song,* Prol. 2.4). Accordingly, Crouzel argues that the agent of learning is the spiritual senses, because they are what allow the soul to apprehend the spiritual sense of Scripture and thereby learn from the Logos (Crouzel 1989, 125). I have already argued that the "process" here is vanishingly slim: it amounts to an equal sign. And from the anthropological perspective, where do the spiritual senses fit in the person? If they are of the inner person, what relation or contact do they have with the outer? We can perhaps agree with Crouzel, who locates the spiritual senses in the upper, spiritual part of the soul. But to "renew" the soul, it seems they must in some way "reach across" from upper to lower, or straddle its spiritual and material parts.

Origen suggests that soul, body, and the spiritual senses are binding intermediaries between matter and spirit. But in each case he simply relocates the fundamental divide within the intermediary, leaving himself with the same problem. Like many Platonist thinkers, he assumes that naming a middle term itself reconciles such divisions. Thus *sensus divinus* is bound to the rest of the human person, and therefore its acts affect that whole person. When it knows spirit, it transforms the whole person into spirit. But if we try to work out how this happens, we must again conclude that Origen does not reconcile spirit and matter *per se,* such that the one perfects or transforms the other. Rather, the *sensus divinus* allows the composite human person to shed its materiality and realize its spiritual essence. We have seen that when the *sensus spiritualis* (or *divinus*) is "trained," it does not "renew" or "perfect" the materiality of *sensus carnalis,* it simply knows spiritual things. The *sensus carnalis* is a solidly nonspiritual, separate set of senses, directed to material things and belonging to the lower or "fleshly" soul; it is not made spiritual, but simply sloughed off, leaving *sensus divinus*. When the *sensus divinus* knows spiritual things—i.e., when it "reaches through" the carnal and materialistic language of the Bible to grasp and understand the spiritual meaning (i.e., the teaching of the Logos)—the lower part of the soul does not understand or know. Only spirit knows spirit and only mind knows mind: "God is not comprehended by any being other than that made in the image of that mind" (*Celsus* 7:78). The lower part of the soul is not redirected to immaterial spiritual things, it is assimilated to the higher, spiritual part of the soul as it loses its materiality and becomes spirit. The soul *as a whole* is transformed, but not because materiality is itself perfected or spiritualized; rather, because its materiality is shed. Origen assumes *sensus divinus* is all spirit and in no way material or bodily; and to the extent that we come to know and return to God by means of the *sensus divinus,* he denies a truly significant role to material body.

Body and Matter in Resurrection and Incarnation

Origen's teaching on resurrection shows a very similar pattern. Here again he argues that, as the Logos acts to spiritualize us, matter and body are not perfected, but lost.[28] Just as he attacks materialist exegesis of the Bible, he attacks materialist conceptions of resurrection (Bostock, 323-25). In order to do so and still defend the idea that we are resurrected in the body, he argues that body is not necessarily a physical, material thing. Citing Paul (1 Cor 15:44), he claims that we are resurrected as we "put aside" material body and acquire a "spiritual body" (*soma pneumatikos*) with the same form (*eidos*) (e.g., *Principles* 3.6). He explains this by distinguishing the body from the elements and "qualities" (*qualitates, poiótetes*) that compose it (Bostock; Crouzel 1980, 192-95, 241-46). Body is formed of the four material elements (earth, air, fire, and water). The elements, in turn, are each composed of qualities (dry, wet, and warm, and cold, hardness, and softness). Qualities are themselves immaterial, but they give form to matter and thus to body. "Qualities, when inserted into matter, produce the different kinds of bodies" (*Principles* 2.1.4). More of some kinds of qualities, such as "hardness," yield a material body; more of others, such as "softness," yield a spiritual one. And qualities are given by God, through the Logos.

Because qualities are not material, the body can retain them and still lose its materiality. Therefore, Origen can claim that, when resurrected, "the [material] body puts aside its former qualities and receives better and different ones . . . by the providence of God's will. . . ."; its material and "mortal" qualities are replaced with spiritual, "divine and ethereal" qualities (*Celsus* 3.41; quoted in Bostock, 330). In this way the material body is progressively replaced with a spiritual one (*Principles* 2.3.3, *Celsus* 2.62; Bostock, 333-35). Origen expresses the continuity between these two bodies with the concept "form" (*eidos*). He does not clearly show how form is made of qualities, except to say that qualities give form to matter; presumably they give form to spirit too.[29] But he does imply that qualities are affected by morals, because he makes clear that a virtuous soul has a spiritual body, or as he more often puts it, a spiritual "state" (*statum*) (Bostock, 332). "State" is probably the better term, because the "spiritual body" seems both immaterial and incorporeal. It cannot be touched or seen, it is free of mortality and corruption, and although it has form, it lacks shape and size. It is "a substance perceptible to the mind alone" (*Principles* 4.3.15). Accordingly, Bostock argues that

> it is difficult to avoid the conclusion that when Origen uses the term 'spiritual body' he is attributing to it all the characteristics of spirit and none of the characteristics of body. . . . And although he makes much

> use of the term 'spiritual body', the word 'body' is effectively evacuated of all meaning. Origen's essential emphasis in fact is upon the soul's advance to a spiritual "condition" and quality—in statum qualitatemque spiritualem" [*Principles* 3.6.6]. (Bostock, 332-34; see also 335-36)

So when Origen explains resurrection, just as when he discusses the structure of the human person, he posits terms to mediate between matter and spirit: in this case, "quality" and "body." But "qualities" are themselves nonmaterial, and he assimilates "body" to spirit. Indeed, his explanation of the resurrection seems to ascribe reality only to spirit and to drain it from matter (Bostock, 324).

It might be claimed that, even if Origen believes material body must be shed for the person to be spiritualized and know God, he also teaches that this can only happen by means of the material world. The material world and material bodies are a divine creation that provide a way for fallen intellects to learn from the Logos, to be spiritualized, and to return to God (e.g., *Celsus* 4.54-57). However we evaluate these statements, his doctrine of the spiritual senses, at least, undermines them. The "learning" and "knowledge" that cause the soul to "advance to a spiritual condition and quality" seem to have nothing to do with material body. They might perhaps be identified with "activating" the spiritual part of the person or with grace. But it is at least clear that Origen's fundamental assumption is that spirit draws to itself spirit, not matter.

One might think that Origen would ascribe more value to material body, because he accepts that the Logos, a spirit, was joined to material body in the person of Jesus.[30] However, although he does not in any way deny Christ's material bodiliness, he is not very interested in it. He doesn't dissolve the physical Christ into the spiritual Logos, but he identifies the humanity of Christ with the soul of Jesus, not his embodied life of suffering and sacrifice. He emphasizes that we know and can return to God through the "names" (*epinoia*) of Christ, especially Wisdom (*sophia, sapientia*) and Logos, rather than emphasizing the bodily work of the incarnate Jesus, crucified and resurrected. This is perhaps why his concept of the spiritual senses is so thoroughly spiritual. The parallel is certainly striking. Origen stresses the teaching function of Christ. Jesus the teacher manifested the eternal Logos, the "image" and perfect expression of God and his first creation (*Principles* 1.2, 2.6.1). The Logos was already present in the created world and the Jewish scriptures, but its presence there was obscure to fallen intellects, who were clothed in and preoccupied by material bodies (*Principles* 2.6.3). Therefore, the Logos took on a body so that it could be a medium for them to learn and be led back through material creation to their origin in immaterial spirit.[31]

I will not belabor how precisely Origen explains the union of Logos and flesh, but simply note that he finds it more interesting than the work of the physical Christ. What is significant for our purposes is his teaching about how the Incarnation made it possible for people to learn from the Logos. The coming of Jesus removed the veil that obscured the Logos within Moses' law; Jesus' actions and words teach "the mind of Christ" (*nous hòn Christon*; *sensus domini*): that is, the Logos (*Principles* 4.1.6; 4.2.3). Because he is material body, and fallen humans understand God by means of intellect, the revelation of the physical Christ "consists in pointing out or indicating God rather than showing him." The basic function of the Jesus Christ, therefore, was to reveal that the Logos was present in the Hebrew law and in the order of the world—and in his actions and words, which direct humanity to a nonobvious spiritual referent (Harl 1958, 340-45). His actions and words teach and introduce the more exalted spiritual teaching, or wisdom, of the Logos.

Origen thinks the physical humanity of Christ is absolutely necessary as a medium that allows human beings to attain the spiritual realm, not least because Jesus' actions founded and gave form to cultic practices and an ascetic way of life that lead people away from and point beyond bodily attachments (Harl 1958, 344-45). But Origen emphasizes the physical man Jesus less than the divine, unchanging Logos. The Logos is the all-important mediator and revealer of God; the Incarnation is most important because it draws our attention to the Logos. But essentially (which in Origen's cosmology means at the beginning and at the end) the spiritual person needs nothing somatic to mediate between it and the Logos, because they are both spirit. At the final *apokatastasis,* all created things will return to Spirit and all materiality will be left behind, including the text of Scripture (*Principles* 2.3.3).[32] Similarly, in this life the spiritually perfect need neither the "bodily" sense or letter of the Scriptures *nor the text itself,* "for to the spiritual person the eternal and spiritual gospel (i.e., Christ) 'presents clearly the mysteries Christ spoke and the enigmas he acted.'" Likewise the spiritual person does not need the bodily Jesus Christ, who does not so much reveal the Logos as veil it.[33] Origen treats almost everything as if it points onward and upward to the God who is spirit. He subsumes the historical actions of the physical Christ into the text of the Bible, effectively treating them as "texts" that, properly interpreted, lead on to the divine Logos (e.g., *Principles* 1.2.7).

In Origen's formulation, *sensus divinus* indicates both the result of proper biblical exegesis (a properly "spiritual" interpretation for dangerously anthropomorphic and corporeal biblical language) and the part of the person that performs that exegesis, to which such biblical language refers: namely, the intellect. There is nothing material or bodily about it. If the spiritual senses form a "bridge" between the inner and the outer person, they are at best the ghost of a bridge because Origen assimilates them so thoroughly to the inner man and to spirit alone. Because the *sensus divinus* operates by way of likeness, and because Origen identifies it so closely with the spiritual, immaterial aspect of the human person, *sensus divinus* fails to bridge the rift between material body and spiritual God. Origen's concept os *sensus divinus* suggests that he correlates all spirit and assumes that spirit gathers its own to itself, with little help from the body. Perhaps paradoxically, it also seems to emphasize the importance of grace. A person can return to God because each is, essentially, spirit, as God is. Spirit is "activated" in the person such that they know spirit only through the grace given by the Logos. A person is "made spiritual" by the Logos—and this happens in the process of interpreting the Bible. The Bible is the only place, in effect, where a person can detect the presence of the Logos, and they do so by means of the mind or intellect.

LATER DUALIST CONCEPTS OF THE SPIRITUAL SENSES

I have gone into significant detail about Origen's concept of the spiritual senses partly because other modern scholars have, and I do not entirely agree with aspects of their interpretations. Specifically, I think many underemphasize the anthropological ideas that his concept addresses. Also, many modern scholars have used Origen's concept as the paradigm or conceptual map to understand later concepts of the spiritual senses and the conceptual model with which to interpret later use of sensory language, especially in mystical texts. This has led them to misinterpret the sensory language in some of those texts and to overlook a nonexplicit concept of spiritual senses that that usage implies. On the other hand, some later theologians did discuss the concept of the spiritual senses explicitly, and their conclusions are, in fact, fundamentally similar to Origen's. I want now to outline, much more briefly, their versions of the dualist concept of the spiritual senses.

It was a thousand years before anyone in the Latin West developed a concept of spiritual senses as highly articulated and explicit as Origen's,[34] but in the thirteenth century, several scholastic theologians did so in quick succession. One reason is that a short passage of Origen's that mentions the idea was preserved in the *Glossa ordinaria;* this encouraged university-trained theologians to discuss it, because commenting on the Gloss was part of their curriculum.[35] The idea also seems to have been in the air; more precisely, the questions it addressed were. They and it were already being discussed by monks in the twelfth century—most prominently, by Bernard of Clairvaux and his friend William of St. Thierry, who were partly inspired by Origen's idea (Latin translations of some of Origen's writings had been preserved in monastic libraries) but were more directly concerned with the basic questions that it addressed: in particular, how the embodied, sensing, experienced self can know and achieve oneness with God. I shall argue in the next chapter that Bernard developed answers significantly different from Origen's.

However, the scholastics did not. Nor, for that matter, had Augustine and Gregory the Great, whose writings were the main witnesses to the idea in Latin from the intervening centuries. All of these writers produced versions of the concept that were dualist and, in many cases, intellectualist as well. Augustine's scattered references to the spiritual senses (along with the somewhat more frequent references of Gregory the Great) are significant because they preserved a vestige of the idea through the early and central middle ages in the West, when no one seems to have been interested in it, and because that vestige bore the footprint of Origen's dualist conception, reinforced by a rhetoric of sensation that strongly favored language of sight and hearing. The thirteenth-century scholastics who discuss the idea in some detail—William of Auxerre, William of Auvergne, Alexander of Hales, Albert the Great, and Bonaventure—are more explicit. They posit two independent sets of senses, rooted in a dualist anthropology and matched by a rhetoric of sensation that avoids or explains away overly somatic referents of sensory language. I mention Bonaventure because he, too, was a university-trained theologian and because he knew some of these other men's ideas on the subject. However, his own treatment of the spiritual senses, although basically dualist, has significant elements of a less dualist conception. Therefore, I will discuss his ideas in a later chapter.

Augustine seems to have known of Origen's doctrine of the spiritual senses and touches on the concept in several places.[36] These are little more than brief allusions. He does not develop the concept as a way to address questions of theological anthropology, much less as a concept that helps him articulate union with God. To the degree that he allows for them at all, Augustine subsumes the spiritual senses within the divine illumination that guides the mind to the truth and the grace that draws the will to God. His rhetoric of sensation, dominated by language of sight and hearing, powerfully reinforces this conception.

At one point in the *City of God,* Augustine writes that we do not find and know God by means of the corporeal senses, which lead out of our inner selves and away from him, but rather by means of "the sense [singular] of the inner man," which, in sharp contrast, allows the mind illuminated with "spiritual light" to rightly distinguish just from unjust, inner from outer (*City of God* 11.27). The two other main allusions to the spiritual senses (plural) are in *Confessions,* where he suggests they are aspects of the love that draws us to God. One builds on the premise that love is led and stimulated by beauty, and beauty is perceived through the senses (*Confessions* 10.6). Accordingly, Augustine contrasts the objects of the corporeal senses, which are limited in time and space, with the objects of the spiritual senses, which are not. His concern is not *how* we love and are drawn to God—that is, not "the spiritual senses" themselves—but "what I love when I love my God"—namely, the attributes that God makes available to the will in the illumination of grace. In the other passage, Augustine also speaks of God's attraction in sensory terms by contrasting outer and inner objects of sense and of love (*Confessions* 10.27-10.28).[37] Here he makes clear these are only the beginnings of the illumination by which God teaches (a major theme that continues through Book 12). He who is light illumines the mind, granting it the light of truth, which is a criterion by which to quest through the temptations of worldly things, from sensory pleasure to mental pride and complacency, and rightly interpret words and signs to find God beyond them.

In all these places, Augustine, to the extent that he allows for spiritual senses, identifies them with the sight of the mind; the sight of the mind, in turn, he subordinates to the illumination conferred by God. This is accords with his powerful rhetoric of sensation, which (unlike his remarks on the spiritual senses) profoundly influenced both the doctrines and the language of later theologians. Augustine articulates his theology using language based on all five senses, but language of hearing and seeing predominates overwhelmingly, which only continues the well-established tendency of Hellenistic and early Christian thinkers, who considered such language of a piece with the intricate metaphysics of Word and Light. David Chidester has argued that Augustine (like others) assumes that sensory language refers to the operations of the bodily senses and implies relations between subject and object that correspond to those required by the relevant sense, and that he plays on these implied relations to articulate his theological arguments.[38] If we go a step further and consider how this rhetoric assumes a concept of the spiritual senses, I think we must conclude that Augustine's rhetoric of sensation reproduces the dualism and intellectualism of "Origenist" concept.[39]

In his discussions of how we come to know God and approach union with him, Augustine only infrequently uses language of the "lower" senses, i.e., taste and touch. Like Origen, he considers the senses of taste and touch "bodily" and seems to recognize that referring to them implies immediate contact. There are some exceptions that possibly hint otherwise. Two good examples that seem significant with reference to the mystical language of sensation (i.e., sensory language applied to union with God) are a phrase Augustine uses to characterize ecstasy, "I passed beyond myself that I might touch him," and his account in *Confessions* of the "vision" at Ostia (*Homilies on John* 20.11 and *Confessions* 9.10; discussed in McGinn 1991, 238-41, 253). In the passage in *Confessions,* he writes that after he and his mother speak together for a while, "attingimus aeternam sapientiam." We can perhaps consider *attingimus* an appeal to touch; however, in the next sentence he describes it as a "momentum intelligentiae," which would seem to assimilate it to intellect, and so some aspect of the *visio* that Augustine parses to establish the relation of intellect to God. Likewise the phrase in *Homilies on John* is explicitly a gloss on ecstatic *vision* beyond the realm of body.

Thus Augustine—unlike, as we shall see, Bernard and Hadewijch—does not appeal to the sense of touch to suggest an immediate relation based on an integral anthropology of soul and body. He rejects the idea of immediate contact carried by language of taste and touch, *except* insofar as it is mediated by the Word and the Light that God is. He assumes that this mediated contact is an affair of learning and knowing, which, like Origen and so many others, he associates with light and sight and hearing. Of course, Augustine conceives of learning and knowing differently than Origen; he does not emphasize the essential bond between intellect, spirit, and the God who is pure spirit, which seems to be realized when it is recognized and nonspirit is purged. Augustine emphasizes the metaphysical *gap* and the difference between God and the human person; he also assigns a much greater importance to what he calls "will" and "memory," in the person, and to the help that God gives, which alone allows us to turn to him, find him, and learn from him. As Chidester makes clear, Augustine refers to the mediated senses of sight and hearing to articulate this group of ideas. Because these senses are mediated, references to sight and hearing allow him to articulate how it is we know, are influenced by, and are drawn to God, while still emphasizing that God is fully transcendent and beyond our (unaided) reach. Humanity is fallen, stuck in the realm of time, change and material embodiment, and we have access to God insofar as our higher faculties can receive the Word and the Light that is God. Thus Augustine tends to identify the human self who is saved with these higher powers of the soul, which merely "uses" a body. Although he teaches that the body, because created by God, is fundamentally good, he consistently emphasizes that "we"—that is, *souls*—are led astray by the body and the bodily senses, which misdirect our attention and our love to worldly, material things. He brings up the spiritual senses primarily to reinforce this point; and he then assimilates them to the "illuminated" mind, and with the graced will, which by means of love is drawn and redirected to God.

Augustine associates learning and knowing, the senses of sight and hearing, and the language that refers to these senses, all very closely with the more deeply "inward" aspects of the soul, which bear the image of God and are the most distinct from body and bodily knowing. In these respects, Augustine's rhetoric of sight and hearing assumes the same basically dualist and intellectualist concept of spiritual senses pioneered by Origen.

One of the main vectors for Augustine's ideas in the first half of the middle ages, especially among monks, was the writings of Gregory the Great, and it is worth noting in passing that Gregory says more about the spiritual senses without adding much to Augustine's assumptions.[40] If anything, he emphasizes the fallenness of body more strongly and, therefore, the stronger potential of the corporeal senses to lead us astray. Gregory also emphasizes that before the Fall, Adam perceived spiritually and directly, without the distorting and interposition of body and bodily senses. He does mention the concept more often. Also like Augustine (and most other medieval theologians), he strongly favors language of sight, light, and hearing: one can find dozens of instances of phrases such as *oculus cordis, oculus animae, oculus mentis,* and *aures cordis.* There are a mere handful of parallel references to *os, gustus, tangere,* and related terms.[41] In contrast, we shall see in the next chapter that one of Gregory's closest students, Bernard of Clairvaux, happily reaches for just this latter somatic language to talk about the worthy monk's immediate relation to God.

In his study of the doctrine of the spiritual senses, Rahner argues that thirteenth-century scholastics—other than Bonaventure—identify the spiritual senses with the intellect, as had Origen (Rahner 1933). I agree, and would add that they show much the same concern to mark a sharp line between spiritual and corporeal senses, based on their defense of the distinction between spirit and body.[42] On the other hand, they draw the line of distinction closer to God than Origen did, partly because they articulate it in somewhat different terms. First, they were influenced by Augustine's ideas. They duly cite his remarks on the subject, but more importantly, they assume a "low" anthropology and its partner, a strong doctrine of grace. Thus they identify the spiritual senses with acts of graced intellect and even directly with the gifts of grace. They regard intellect as a *potential* point of contact with God, which must be perfected by grace. Although we retain the image of God, especially insofar as we are rational creatures with free choice of the will, we retain no fundamental bond with God that allows us to know or be one with him without his help. We have through sin lost our likeness to him; crucially, this affects our ability to learn and know.

We can contrast this with Origen's "high," optimistic anthropology, which allows for a more self-propelled divinization than the scholastics or Augustine ever did. The scholastics do not identify intellect so closely with "spirit" that of itself learns from and knows the Word or God. This brings us to a second difference from Origen: the scholastics are not very interested in the exegetical roots of the idea of the spiritual senses, but assume it is a concept that has to do with knowledge and the structure of the soul (psychology). And, of course, they discuss this structure using the Aristotelian ideas that had caught the imaginations of intellectuals in the twelfth and thirteenth centuries. Although the details are complicated, it can be said broadly that they defend epistemologies that root knowledge of God and spiritual things in the higher powers of the soul (intellect and will), which know God insofar as they are distinct from the lower powers of the soul that connect it to material body (sensation and imagination).

The earliest scholastic to discuss the spiritual senses at any length was William of Auxerre (d. 1231). He reduces the spiritual senses to one, and identifies them with the intellect that is "perfected" by the theological virtue of faith—or perhaps faith *is* the spiritual senses; he is not entirely clear. Faith follows on grace, so the spiritual senses seem to be acts of faith.

We enjoy (*fruimur*) God only by faith,[43] not by charity, which merely moves us toward this enjoyment and conjunction; but the joining of like with like is "by cognition, for intellect alone is the image of God" (*Summa aurea* lib. 4, tract. 18, q. 3, art. 2) [in cognitione, quia solus intellectus est ymago Dei]. Unlike the five corporeal senses, which are five in essence, the spiritual senses are one in essence: they are intellect. They are—it is—perfected not by five virtues, nor even by the three theological virtues (faith, charity and hope), but only by the one, faith. "The spiritual sense is one, namely intellect, and one virtue alone perfects it, namely faith" (*Summa aurea,* lib. 4, tract. 18, q. 3, art. 3). [Unus solus sensus spiritualis, scilicet intellectus, et una sola virtus perficit sensum spiritualem, scilicet fides]. In heaven the spiritual sense is one operation in *visio Dei.* In this life, it is both one and five: one power, habit, and principle of perception, namely faith (*virtus, habitus, ratio percipiendi*), but fivefold in operation, according to its object. Faith sees, hears, and loves Jesus spiritually, and knows his work for us, smells his good odor, and by meditation eats, chews, and tastes his sweetness (*Summa aurea,* lib. 4, tract. 18, q. 3, art. 4).

It is not clear whether these "five" spiritual senses are acts of the single spiritual sense that is intellect or acts of faith "within" that intellect. But it is clear that the spiritual senses are acquired from God through grace. That is, they are not something *of us* that allows us to know God but that knowledge itself, given to us by God. This is essentially a version of Origen's idea that the spiritual senses are distinct from body and matter: William regards the spiritual senses as something of the spiritual realm, but articulates the point in terms of grace.

Much the same concept is outlined by William of Auvergne (or of Paris; c. 1180-1249). On the face of it, he seems to distinguish some spiritual senses from intellect. He assigns only sight and hearing to intellect; smell and taste he assigns to the will and calls *motivae virtutes* (this is in a text on the virtues).[44] Thus he reproduces the traditional ranking of senses, with sight at the top, and the base, bodily senses at the bottom. On the other hand, citing Bernard of Clairvaux, William specifically connects spiritual taste with the gift of wisdom (*De virtutibus* cap. 11). (I shall discuss Bernard's formulation in the next chapter.) To the degree that he assumes wisdom searches spiritual things by "experience" more thoroughly than intellectual knowing can, this implies that he considers spiritual taste the highest of the spiritual senses. But in fact the "experience" of wisdom that the will "tastes" is a grace that stimulates us to know God more intimately, and so subordinate to proper knowledge of God, which comes by hearing and sight—by intellect.

However, these distinctions among the spiritual senses are slight. The Parisian doctor presents the spiritual senses as a vague general faculty that receives, perceives, and distinguishes spiritual "things" in the broadest sense. He writes that, by means of the spiritual senses, a person senses spiritual passions (e.g., sadness and joy), spiritual knowledge and doubt, visions and illuminations (at least the prophets did), and can receive theology or divine philosophy.[45] This grab-bag of incommensurate spiritual "objects" perhaps helps him distinguish them from the corporal senses, but mainly allows him to emphasize the distinction between worldly and spiritual things as general classes. His vague conception of the spiritual senses accords with his ideas on epistemology. He posits an "intuition of the intelligible [i.e., "spiritual things"] which duplicates that of the sensible"—essentially a version of Augustinian illumination.[46] Knowing is not by means of faculties, but directly by means of the "essence" of the soul, which is God in the soul. The creator is the eternal model by which everything is visible and knowable "joined to, and very present in, human intellects . . ." (Gilson 1980, 257, quoting William of Auvergne's *De anima*). Intellect, William, argues, is "the very essence of the soul exercising its knowing activity" (Gilson 1980, 256). Soul is absolutely simple: "essence is the immediate cause of its operations of intelligence and of will; no distinct faculty of the soul is interposed between it and the operations it accomplishes." So too for the spiritual senses. They disappear into the essence of the soul that is intellect.

Alexander of Hales (c. 1186-1245) says rather more, but much the same.[47] Under the heading of "cognitive reason" (*vis rationalis cognitiva*), he classes the spiritual senses as "intellective acts" (*actus intellectivus*) (*Summa,* memb. 3, preface). The spiritual senses have a spiritual nature; spirit is above mind, and it is called intellect as intellect is called spirit (*Summa,* memb. 3, cap. 1). The spiritual senses are five not because they are distinct powers of the soul nor (as the two Williams have it) because their objects differ but because their "modes of comprehension" differ (*Summa,* memb. 3, cap. 2). [Plures dicuntur sensus, . . . non quia obiecta sunt plura vel quia plures potentiae, sed ratione modorum comprehendendi diversificantur]. Different aspects of one object are apprehended by different senses in different ways, which operate from one unitive sensible power in the soul. (This is the *sensus communis,* the mediating faculty that "translates" and "passes along" the data of the sense organs to the imagination and memory, then to intellect.) Alexander does not specify what these "modes" might be with respect to the interior senses (*Summa,* memb. 3, cap. 2). (He cites Augustine, *Confessions* 10.6, which was a major *locus* for scholastics' discussions of the spiritual senses but far too vague to serve as an explanation.)

We are left with the statement that the spiritual senses are intellective acts that operate through the *sensus communis.* Perhaps they are special acts of the *sensus communis;* or perhaps they are the *sensus communis* itself. In the course of his treatment Alexander slips from discussing five spiritual senses (*sensus spirituales*) to discussing a more general concept of *sensus interior,* which seems to be identical with the *sensus communis*

(*Summa,* memb. 3, cap. 2). He explains that *sensus interior/communis* is partly inseparable from body, insofar as it pertains to the sensible aspect of the soul, namely the imagination, which mediates sense images to intellect. But in other respects the *sensus interior* is separable and independent of body, insofar as it pertains to the properly rational aspect of the soul, in the intellect proper and the *affectus,* the soul's power of desire and love (*Summa,* memb. 3, cap. 3). This aspect of *sensus interior* is present to disembodied souls and angels, who are directed to "intelligible things" (*intelligibilia*), which are of course entirely spiritual and in no way bodily (*Summa,* memb. 3, cap. 4). If Alexander means *sensus interior* as a synonym for *sensus spirituales,* then he seems, like Origen, to want it or them to span or "reach across" the distinction between body and spirit. But this is only a hint. His main point is clearly that they are intellective acts, and are related to the corporeal senses only by name.

Albert the Great (1206-1280) devotes more attention to the spiritual senses than does any other scholastic except Bonaventure. His treatment stands out both for its subtlety and for explicitly addressing the issue of "mystical" knowledge—that is, knowledge of the hidden things of God. In his commentary on Dionysius' *Celestial Hierarchies,* Albert explicitly assigns the spiritual senses to the intellect, which allows us to grasp God as Truth and as Good and so ascend to him.[48] On the other hand, in his commentary on the *Sentences* he (like William of Auvergne) assigns only spiritual sight and hearing to the intellect, locating taste and touch to the *affectus* (and so to the will, with which the *affectus* is closely associated). The distinction corresponds to a difference of object: intellect, by spiritual sight, hearing, and smell, perceives God as Truth; *affectus,* by spiritual taste and touch, grasps God as Good. But even when introducing the *affectus* Albert insists that all the spiritual senses are forms of *cognitio,* and at the end of this discussion he identifies the different spiritual senses as acts of the "speculative" intellect, which he distinguishes from "possible" intellect (*In III Sent.*, dist. 13, A, art. 4).[49]

Albert's treatment is remarkable, both because he was the first to explain the concept with explicit reference to Dionysian mystical theology and because he was the first scholastic to make any serious attempt to distinguish the spiritual senses according to their different ways of operating (although, as we shall see, the monk Bernard of Clairvaux had done that before him). Like others, Albert distinguishes the spiritual senses by itemizing objects for them; however, influenced by Dionysius, he chooses only two, and they are the highest attributes or names of God in himself, Good and Truth. And Albert's goal is to distinguish between the differing operations of the spiritual senses. According to the treatment in his *Commentary on the Sentences,* God as Good is known in two ways: in the mode of touch, by conjunction, which is "by experience"; and in the mode of taste (*sapor*), by mixing. To counter the objection that these senses do not apprehend but are merely influenced passively, Albert cites Dionysius: spiritual taste and touch are experiential knowing (*cognitio experimentalis*). Truth is known in two ways as well, with a subdivision, making three. The lowest is *in ratione boni,* in the mode of smell, when the good "evaporates" a waft of the good from itself, so that truth is known from a distance, as it were. *In ratione veri,* we known truth as truth, either in a sign or in itself (*aut in signo, aut in se*). In a sign it is perceived in the mode of hearing, which grasps the form of a thing in a sign. In itself, truth is grasped in the mode of sight, directly, "which is in the form of a thing *per se*" (*III Sent.*, dist. 13, A, art.4; Albert says the same in *De cael. hier.* 15.5, although there he discusses the different operations at greater length).

It is striking that Albert distinguishes the senses partly according to spatial relations. He ranks sight and hearing as primary, precisely because they grasp Truth in itself, from within (*per medium intrinsecum* [*De cael. hier.* 15.5]). Taste and touch grasp Good by something like direct contact, but it is only from without (*ab extra* [*III Sent.*, dist. 13, A, art.4]; *per medium extrinsecum* [*De cael. hier.* 15.5]).[50] Albert seems to consider this experiential knowledge less probing than the sight that is intellectual knowing proper. As we shall see in the following chapters, this is exactly opposite to the assumptions of Bernard and Hadewijch. And it is highly significant, because even if Albert means to imply that intellectual and "experiential" knowledge of God complement one another, the way he ranks them underlines his premise that the spiritual senses are cognitive acts of the intellect alone, not anything like a unified sensorium that belongs to the whole person. The basic dualism of Origen's concept is intact.[51]

The treatments of these scholastics are not thorough and do not agree at all points. Yet, working with an Aristotelian faculty psychology, they all agree that the spiritual senses are aspects of or acts of intellect, and agree on the sharp distinction between them and the corporeal senses. Two of them, William of Auvergne and Albert the Great, perhaps mitigate this insofar as they root some of the spiritual senses in the *affectus* (the soul's power of desire and love, which is not quite a faculty of itself, but associated closely with the will). Similarly, Albert associates spiritual taste and touch with "experience." Yet he and they regard still subordinate this knowing to the knowledge gained by the intellect that "sees."

CHAPTER 3

Bernard of Clairvaux:

Spiritual Sensation

The Cistercian monk Bernard of Clairvaux (1090-1153) knew about the idea of the spiritual senses from reading Augustine and Gregory the Great and also knew Origen's discussion.[1] In a few places, he discusses the spiritual senses explicitly and suggests that they are a "second set" of senses. More significant, however, is the way Bernard uses sensory language extensively to discuss how we know and achieve union with God.[2] Of course, building on a long tradition in Christian writings and Latin literature (Augustine and Gregory the Great, in particular), he frequently uses language of sight and hearing; he also mentions the traditional view that these senses are the most appropriate to God. But he also uses language of taste and touch to articulate the immediacy of union with God. He does not explain it away as language that actually refers to the intellect that "sees" and knows things of the spirit. Rather, Bernard's usage suggests that the human person has a single sensorium that can be directed both to material and bodily things and to spiritual and divine things.[3] This concept of "spiritual sensation" follows from his assumptions about the structure of the human person; in particular, Bernard often does not sharply divide its spiritual and the bodily aspects. His concept of the human person has dualist elements, but it is in many respects integral: the human person who loves, knows, and achieves union with God is a unified individual, soul and body. We can trace these assumptions through many strands of Bernard's theology, especially his ideas about anthropology and epistemology. Two are central: his emphasis on the importance of the incarnation of Christ and his emphasis on the need for the individual monk's "experience" (*experientia*), which, he clearly assumes, is gained only by way of the body.[4] "Experience," in turn, is central to exegesis, in particular exegesis of the Song of Songs, which Bernard calls "the book of experience." Bernard, like the men I have discussed so far, interprets the sensory language of the Bible to refer to spiritual things—but only partly. For the Song of Songs sings of experience, and, therefore, it sings of the body. The sensory language in the Song (and the Bible as a whole) is not simply to be explained away; Bernard assumes it refers both to body *and* to spiritual things; this is perhaps why he thinks it is perfectly suitable to talk about the intimate union of the self and God.

INCARNATION, EXPERIENCE, AND THE SONG OF SONGS

To grasp how and why Bernard uses language of touch and taste, we must take a quick tour through a few of his theological ideas. I will group these around his understanding of the concept *experientia* and the significance of the Song of Songs. Bernard thinks that human beings are enslaved to the material world and fail to perceive God in it: we are stuck in carnality, and love spiritual things carnally. Although the body impedes a person from renewing the *imago Dei* and conforming themself to God, it is also what allows the person to gain access to God, because it was as the physical man Jesus Christ that God revealed himself and provided human beings a way to know, love, and become one with him. We find and love God not by escaping the material world, but through our individual experience and our love of the human, bodily Christ. That is, corporeality is both the condition of fallen humanity and the condition for fallen humanity to know and love spiritual things. Human beings cannot embrace the spirit unless they first embrace the body; we need a relation to body because we understand nothing else. God "took on flesh for those who know the flesh, to teach them to taste and know the Spirit." [Obtulit carnem sapientibus carnem, per quam discerent sapere et spiritum.][5] If we first love the physical being of the God-man who came as a man in history, suffered, and died,[6] then we love the historical, physical manifestation of God's attributes in the virtues of Christ, and then we can come to love him spiritually as the redeemer who was resurrected and ascended bodily to heaven. We thereby "ascend" from love of God as a carnal man who suffered and died to love of the ascended Christ who is seated with the Father in heaven. In Bernard's view, the perfection of this love is *unitas spiritus,* in which the soul and the divine Bridegroom are united in love and mutual willing. He finds the model for this union in the Incarnation (a point I will return to in a moment).

Bernard understands biblical exegesis according to the same pattern. Christ is present to humanity in both the historical and the metaphorical senses of Scripture (*SC* 6 and 20). Just as we need the physical Christ to restore our misdirected love, so we need anthropomorphic, physical images of God to come to love the God who is spirit with a spiritual love (*SC* 4). As physical creatures, we need the physical to attain pure spirit (*SC* 4 and 5). Here we find a difference between Bernard and Origen that bears on the way they interpret the Bible's sensory language. Bernard stresses that the object of exegesis is the incarnate Christ; Origen stresses that it is the eternal Logos. Origen seems most interested in the words of the Bible as points of access that point "above"—portals for the mind to find the realm of pure Spirit. For Bernard, on the other hand, we cannot know or love God without the historical, physical Christ present to us in the Bible. Christ's humanity and life in the body makes possible for us to realize our likeness to God. The God who is spirit would be inaccessible to fallen humanity without Christ's body and ours. Corporeality is not the problem, but the solution to our misdirected love.

Bernard also teaches that the love of the physical Christ must yield to a spiritual love; but spiritual love has a bodily component. First, such spiritual love is still by way of Christ: the ascended Christ. This implies a relation mediated by the physical body, because Christ rose from the dead and ascended in the body, not as pure spirit. Second, Bernard emphasizes that "it is not our bodies but our sins that stand in the way" of union with God (*SC* 56.3; *SBO* 2:116) [obstant, non corpora, sed peccata]. Accordingly, at the final judgement the *whole* person will be raised and enjoy union with God: not just the immortal soul, but soul and body. Moreover, he writes that the soul's joy, repose and "inebriation" is augmented and perfected at the final resurrection because the soul is then rejoined to the body (*Dil.* 10.27-11.33). Soul and body advance to union together.[7] Therefore, what is conformed to God is not just the soul, but the whole person. The spiritual marriage between the bride and the divine Bridegroom is a conformity of wills and unity of spirit, which is "experienced" in the whole person, soul and body. The integration of soul and body is also implied by Bernard's insistence that union with God is necessarily manifest in the world as love of neighbor, a life of virtue and *caritas* that is evident and perceptible as "experience."

Bernard also binds "experience" to exegesis. In the *Sermon 3* on the Song of Songs, he famously declares, "today we read in the book of experience" (*SC* 3.1; *SBO* 1:14) [Hodie legimus in libro experientiae]. The "book of experience" is the Song, which tells of the individual soul's encounter with God in love, and it is also the book written on the heart—that is, on the affect and the memory of the contemplative monk who seeks God in love. The two books of experience reflect each other. When Bernard tells his monastic audience that they will "read in the book of experience," he in effect says that they will measure and interpret the Song by their experience, even as they measure and interpret their experience by it. They will set themselves between "the book of experience" and "the book of Scripture"; or to put it another way, they will set themselves within the Song, because it is the scriptural account of experience.[8]

Bernard's point is that experience is a moral teacher that leads us to God. It is by experience that we know that we are sinful and that God is merciful, when we feel compunction and holy fear, and confidence in the benefactions of God; it is by experience that we come to know and practice charity; and it is the experienced exegete who can be taught by the

Song—the book of experience—both intellectually and sensorially, and who can ascend from the historical sense and carnal love to spiritual sense and spiritual love. The book of experience, like the book of scripture, is both pedagogical and anagogic.[9]

Yet if experience bears us "upward" to the realm of spirit, it is not therefore opposed to or distinct from body: Bernard thinks experience is intrinsic to human life in a body. In a fascinatingly tortured section of *The Steps of Humility and Pride,* he asserts that Christ experienced and learned from experience because he was a physical man.[10] He instantly confronts a theological problem: Christ is God and so knows all and needs nothing; how then can he learn through experience, as humans do? Bernard struggles with this question for several paragraphs and fails to come up with a satisfactory answer, and in the process he reveals his central assumption: that to experience is to have—to *be*—a body. In a similar (less thorough) discussion in *Sermon 56* on the Song, he summarizes his conclusion as follows:

> On being made man, therefore, he [Christ] used our bodily feelings and senses as openings or windows, so that he would know by experience the miseries of men and might become merciful. These were things he already knew but in a different way. . . . Do you see him becoming what he [already] was, and learning what he [already] knew, seeking in our midst openings and windows by which to search more attentively into our misfortunes? (*SC* 56.1; *SBO* 2:114-15)[11]
>
> [Humanis ergo affectionibus corporeis pro foraminibus usus est et fenestris, ut miserias hominum homo factus experimento sciret, et misericors fieret. Sciebat et ante, sed aliter. Videsne factum esse quod erat, et quod noverat didicisse, et sibi apud nos quaesisse rimas et fenestras, per quas calamitates nostras diligentius exploraret?]

Bernard argues that Christ did not so much learn new things by experience, but experienced "proofs" of what he already knows in "another way." Again, this attempted solution reveals Bernard's basic assumption: *bodies* experience.

Therefore, when he emphasizes that human beings find God in their experience, he emphasizes that they find God through their bodies. Experience has a mixed character, both spiritual and somatic. It is not merely something that happens to the body, but it includes those aspects of physical existence in the world that lead us to love the embodied man Jesus Christ, which allows us to achieve *unitas spiritus* with the ascended Christ. By experience we realize our love for and likeness to God. Experience is a dynamic program of renewal: humanity "can through order, discipline and obedience, through reform of his physical and spiritual life, imitate a more divine model" (Stock 1976, 227-28). It indicates how we should relate to the created world, including our bodies, in order that we can return from the *regio dissimilitudinis* to God.

Bernard understands our human perspective on this relationship by means of Paul's phrase, "utentes hoc mundo tamquam non utentes" (2 Cor 7:29-31). "Using the world without using it" can renew the *imago Dei* in the physical person. The "outer man" is gradually bound to and conformed to the "inner man" as "experience informs action and action in turn instructs the mind" (Stock 1976, 229). This principle also governs our devotion to Christ. We should love and contemplate his physical nature, his virtues, and his worldly sufferings, but we must recognize them as a step or bridge to true spiritual love for the spiritual God, beyond physicality.

It is clear that Bernard patterns this concept of experience on the Incarnation. The Incarnation models the action that will lead us back to God, because it models all intercourse between the physical and the spiritual. The model does not explain the connection in the manner of a logical proof but indicates the proper course of action in the world while making clear that choosing the path of piety and virtue is not in our purview but in God's. Bernard treats the senses according to this same incarnational pattern. He posits no distinct *second* set of senses or powers by which we know God, as Origen does. He shows no interest in speculative psychology: instead of "the idea of the structure of the soul," Bernard's use of sensory language suggests "the idea of a structure in the domain where the soul exercises its powers and deploys its activity: the diversity is in the objects toward which the soul can orient itself."[12] He teaches that the object and activity of the human senses can be redirected and spiritualized. The corporeal senses must be humbled and controlled so that they may be conformed to the spirit, which happens when, with the help of grace, we "use the senses without using them."

Bernard is willing to articulate how we know God with language of taste and touch and let its somatic implications stand because he assumes we know and can be one with God as whole people, soul and body. He also assumes its spiritual reference is secure because he refers to these senses in the language of the Bible and in the context of biblical exegesis. The Latin Bible provides him with both a vocabulary of sense and the opportunity to expand on it in the act of exegesis, especially exegesis of the Song of Songs.[13] He cites sensory language from all over the Bible—he is particularly fond of Ps 33:9 ("gustate et videte quoniam suavis est deus") and 1 Cor 6:17 ("qui adhaeret Deo unus spiritus est," a clause he interprets as a reference to the immediacy of touch)—but the Song is thickest with language that stimulates him to expound on the meaning of sensory words and phrases. The Song provides Bernard with two of his favorite references to touch: the kiss, from Song 1:1 ("osculetur me osculo oris sui"), and the embrace, from Song 2:6 ("laeva eius sub capite mea, dextra amplectitur me"). The fragrant ointments mentioned in Song 1:3 rouse Bernard to discuss the soul's relation to God in terms of smell; similarly, he expands on hearing and vision of God when discussing Song 2:9-10, which depicts the Bridegroom looking through the lattice and speaking to the bride.[14]

Bernard inherited a Christian tradition of Song interpretation, established by Origen and maintained by monks, that found in the Song an authoritative language of love to model the way Christians relate to God.[15] Origen interpreted the Song as a drama of God's encounter with the Church as a whole and with the individual soul. Although early medieval exegetes were most interested in the ecclesial and eschatological interpretation, by the twelfth century monks were increasingly interested in the individual. They interpreted the Song as a drama or map of the monk's own search for, love for, and "mystical marriage" to Christ.[16] Bernard's influential sermons on the Song of Songs ratified and encouraged this trend. Although Bernard did not ignore the ecclesial interpretation, he put a strong emphasis on the individual aspect. He strongly emphasized that love is the central fact of Christian life, and especially the life of monks. The Song had special importance in the life of the monk because, even as it told of the love between the bridegroom Christ and the Church his bride, it could also teach the monk about his individual quest for God through love (a quest that Bernard assumed was an important part of the work of the Church).

To read the Song as an erotic drama between Christ and the individual soul always risked a dangerously carnal interpretation. Origen warned against this and tried to forestall it, arguing that the Song had a *only* spiritual meaning. Bernard echoes this warning several times: those who are trapped in the flesh and are unworthy should not read or listen to the Song because they easily misinterpret its sensual language (see esp. *SC* 1.3, 1.6, 1.12). He clearly assumes that occasional warnings are enough, because he does not often bother to hold *himself* back. He appropriates the Song's potentially dangerous language with enthusiasm (see Turner, 164-47). One reason Bernard could do this was that, by the twelfth century, the Song was tamed as it had not been for Origen in the third. Its meaning had been moored within a monastic tradition of exegesis that precluded overly carnal interpretations. In the monastic tradition, a carnal reading of the Song would not be a reading of the Song at all, but a fundamental misapprehension that would, in effect, create a new text where the Song of Songs had been. The monastic tradition of Song exegesis anchored the Song's language, including its most dangerously corporeal sensory language, within a realm of spiritual signification. Because the Song is the extreme instance and the exemplary case of highly somatic and erotic language in the Bible, the monastic tradition of Song exegesis that anchored the Song within a spiritual realm of signification also anchored its corporeal language there. Among monks, such language could therefore be used to discuss human knowledge of a spiritual God—or better, the conformity with God that is available in this life. At least Bernard thinks so. Although he does at times insist on the spiritual meaning of somatically charged sensory language, he also accepts and plays with its somatic implications, without fearing that they would drown out that spiritual meaning: the two coexist.

THREE FRAGMENTARY DISCUSSIONS OF THE SPIRITUAL SENSES

I have sketched a few of the theological, anthropological, and interpretive assumptions that govern Bernard's use of sensory language, and I have suggested that his usage assumes a concept of spiritual senses distinct from the dualist and intellectualist family of concepts I outlined in the previous chapter. Before examining some of that language to show how that is the case, I want to touch on three places where Bernard discusses spiritual functions of the senses explicitly: the two sermons collected as 10 and 116 *On Various Topics* (*De diversis*) and the sermon or report of sermon which we have as *Sententia 3:73*.[17] In each he matches and ranks the five canonical senses with a corresponding form of love by which we know God. The treatments are sketchy, but he appears to be talking about a separate set of senses that would fit with the "Origenist," dualist model. On the other hand, he is not that interested in them. He essentially exploits the five senses as a rhetorical motif to elaborate on his main subject, love. Still, these sermons allow us to make three useful observations about Bernard's understanding of the physical and spiritual senses. First, they make it clear that Bernard was aware of the basic idea that the human person has a sense or senses that allow it access to spiritual things. Second, they demonstrate that Bernard's explicit speculations about this concept are subordinate to his interest in love. Third, and perhaps most significant, these sermons show Bernard thinking about the operations of the corporeal senses as models for the soul's relation to God. In this respect, the most significant sermon is *De diversis 10*, where Bernard organizes the senses according to how immediately they perceive their objects. He does so inconsistently, and he does not draw out the implications of this approach. But this treatment strongly suggests the distinction between mediated and immediate senses that seems to influence Bernard's use of sensory metaphors throughout his writings.

In *De diversis 10*, Bernard uses the five senses to organize a discussion of love (*SBO* 6,1:122-24). He writes that, just as the body gains its life and sense(s) (*sensus*) from the soul, so the soul gains life and sense from charity. The "life" of the soul is knowledge of truth, which is possessed partially by natural reason or fully by a spiritual soul that adheres to and is one spirit with the truth. The "sense(s)" of the soul is charity or love. [Est ergo animae vita veritas, sensus caritas.] (*Sensus* here can be taken to mean either "understanding, faculty of understanding," or "sense(s)." It was be more precisely defined by dialecticians, but Bernard uses it ambiguously.)

Just as the soul gives life to the whole body uniformly, so too truth is one in the soul. However, sense is not one, it is fivefold and distributed among the members. So, too, the soul's corresponding "sense" is love or charity, which Bernard parses into five kinds, correlating each with a bodily sense. He ranks sight highest, corresponding to "holy love" [amor sanctus sive devotus] of God, followed in descending order by hearing, smell, taste, and touch, which corresponds to "pious or dutiful love" [amor pius] of parents, which is common to people and animals. Here Bernard follows the traditional order that ranks the senses according to their "nobility": that is, the more "physical" senses are more base, and the more "incorporeal" or "intellectual" senses are more noble.[18] However, he seems to be using the "five senses" as a rhetorical armature for a homily on the loves proper to his monastic audience. He does not offer the kind of supporting analysis we find, for instance, in *On Loving God*; nor does his treatment in this sermon agree with that work.

The most theologically promising point in this sermon is that love is the sense of the soul, but Bernard does not expand on this idea (here or elsewhere).[19] Equally interesting and more developed is the rationale he offers for why each sense agrees with (*convenit*) a kind of love: he correlates the "remoteness" of the objects of the senses with the remoteness of the objects of love. Thus the sense of touch requires the object sensed to be in direct contact with the body, while the sense of smell can detect objects at some distance, and the senses of hearing and sight at still greater distances. So too with love: parents closest, neighbors near, God furthest. Bernard's treatment is somewhat inconsistent. He fails to note that we can smell and hear things which are too far off to see, and does not discuss the fact that tasting too relies on direct contact. (His point here is ambiguous: he asserts that we need love of companions as we need taste—presumably he means we need to *eat*.) In fact he does not use this principle of remoteness at all for the sense of taste or the love that "agrees" with it.

De diversis 116 has somewhat different ranking of the senses and a sketchy discussion of love as the soul's "life and sense." Here Bernard veers closer to the dualist concept that we have two sets of senses that are distinguished by their different types of objects. As in *De diversis 10*, he claims that "just as the body, that is the exterior man, will receive back its life and sense at the time of resurrection, so too will the soul or interior man receive back its life and sense; that is, knowledge and love" [Et sicut corpus, id est exterior homo, in resurrectione sua vitam et sensum recipit, ita et in resurrectione sua vitam et sensum anima, id est interior homo, recipit, id est cognitionem et amorem.] (*Div.* 116). Just as the corporeal senses are five, so the soul's sense of love is divided into five parts. Bernard does not name these parts but claims that each is attracted to one of the five invisible things of God, which it loves on account of certain characteristics which, we can assume, the loving soul perceives or acquires. The soul loves truth as by sight, on account of its freedom (*libertatem*); justice as by taste, on account of its equality (*aequalitatem*); wisdom as by hearing, on account of its sweetness (*suavitatem*); charity as by smell, on account of its virtue (*virtutem*); and eternity (*aeternitas*) as by touch, on account of its security (*securitatem*). We can only assume that he intends the order of the corporeal senses, as he lists them, to correspond to his list of "attractions." That is, because sight, truth and freedom are first in each list, some aspect of love corresponding to sight is attracted to Truth, on account of the freedom Truth grants. Likewise, some aspect of love corresponding to hearing is attracted to Wisdom on account of its "sweetness," or perhaps on account of the sweetness it grants to the soul. Perhaps this is not a safe assumption; the order of the senses is not traditional and in one place is inconsistent, and the surviving text is brief, unclear and perhaps corrupt (*Div.* 116).[20]

Sententia 3:73 seems to be a somewhat longer and more confused version of *De diversis 10*. It is less coherent and even less well integrated with Bernard's basic theological ideas; but like *De diversis 116*, it suggests something like Origen's two sets of senses. It repeats, with slightly different terms, the correlations between the soul's senses (here called "spiritual senses" [*sensus spirituales*]), different forms of love (e.g., *amor spiritualis, amor Dei*) and different objects of love (e.g., *inimicos, Deus*). However, the sermon lacks the psychological hint, made in *De diversis 10*, that "love is the sense of the soul." Nor do we find anything about how close or far the organs of sense are from their objects. Instead, each corporeal sense is appraised according to how much it is "of the body" and how much "of the soul," and this is then likened to how much the corresponding form of love is "of the soul" and how much "of the spirit."[21] This distinction is partially based on the operations of the senses, but it seems more that the characteristics of the senses are dictated by those of the different kinds of love. Thus love of parents is universal and inescapable, just as the sense of touch is. More typically: as the sense of taste is of the body but gives rise to a taste sensation that affects the soul, so the love of companions is of their collective body, but produces their spiritual feelings.

The most potent idea in these sermons is the idea that love is the "sense" of the soul by which we know God. It suggests that the "sense" of love is a grace that recapitulates salvation history: as God went out from himself in the human Christ in order to allow fallen humanity to return to him, so divine love goes out from him into human souls to allow them to return to him in love and thereby "know" him as by a sense. The analogy is clearest in the last part of *Sententia 3:73*, where Bernard invokes the theory that sight operates by means of a ray that proceeds from the eye, strikes an object, then returns to be sensed and be taken up by the memory.[22] Bernard does not develop this idea further. In any case, the notion tends to collapse the "five" spiritual senses into one. In all these sermons he has to stretch to allow for the "fiveness" of sense. Bernard sidesteps part of the problem by finding loves and objects of love that are virtuous and holy but not in God. However, insofar as all these loves are aspects of the one unified love that we have as a gift from God, they are all one. Bernard never really grapples with this issue, I think because he does not see the need: he does not usually assume that humans have a distinct second set of spiritual senses.

THE ORDER OF SENSES

What these fragmentary sermons *do* show us is that Bernard thought about how the different operations of the various senses might model the soul's relation to God. In *De diversis* 10, where he orders the senses according to how remote their objects are from the body and the organ of sense, he suggests distinction between mediated and immediate senses. The distinction accords both with his basic theological ideas and with the way the pattern of sensory language in his writings. Things we know by fragrance are "hidden" and known at a distance (see *SC* 67.7; *SBO* 192-93); odor communicates the goodness of God at one remove, through virtue and prayerful praise (e.g., *SC* 67.4-8; also *SC* 9.6, 9.10). But spiritual things can be "tasted" and "touched" directly. In *Sermon 19* on the Song, Bernard writes that "the odor of the spouse's outpoured oil rouses them [the maidens] to taste and feel that the Lord is sweet [Ps 33:9]" (*SC* 19.7; *SBO* 1:112; see also *Sent. 3:121*). [Habet oleum effusum sponsa, ad cuius illae {adulescentiae} exitantur odorem, gustare et sentire quam suavis est Dominus.] In *Sermon on Psalm 90* ("Qui habitat") he designates the immediate presence of God by the touch of a kiss: "Not content with messengers, even that one who speaks [to you], you pray and hope that he [Christ] might be present to you, not by any kind of medium, but that he kiss you with the kiss of his mouth" (*QH* 15.2; *SBO* 4:476).[23] [Oras et optas, ut nuntiis non contentus, ipse quoque qui loquebatur, adsit, nec te quasi per medium, sed sui oris osculo osculetur.] In *Sermon 67* on the Song, Bernard connects this same itinerary with "setting love in order." We have already heard him lament that the experienced taste by experience what he only smells: "How long shall we smell and not taste, gazing toward our homeland but not grasping it, saluting it from afar with sighs?" (*SC* 50.8; *SBO* 2:83). [Quousque odoramus et non gustamus, prospicientes patriam et non apprehendentes, suspirantes et de longe salutantes?] And the bride of the Song, though she enjoys the exalted taste of contemplation, can only hope for the dawn of a new life in heaven, when she will enjoy the clear and everlasting *visio Dei* (*SC* 48.8).

The distinction between mediated and unmediated senses is not obvious, and some might say it is not worth seeking, because the way Bernard uses sensory language does not fall into any discrete hierarchy or pattern; rather, it is "synaesthetic."[24] It is true that in a number of places Bernard refers to two or more of the bodily senses in a single phrase, clause, or sentence, using words that apply to one sense with reference to another. For example, he writes clauses such as "you will touch me with the mind's eye" (*SC* 28.10; *SBO* 1:199) [tanges oculo mentis]. More broadly, he writes of the soul's intimate encounters with God using metaphors based on more than one sense: in *On Conversion* Bernard explains that the converted soul, once it has heard the word of God and cleaned the inward eyes of the heart, knows an "inward paradise." He describes this paradise with mixed sensory language taken from the Song of Songs: in the enclosed garden with its sealed fountain, the bride hears the voice of the turtledove, smells the sweet scents of blessedness, and tastes the sweet fruit of the apple tree (*Conv.* XIII.25). He calls this a vision of pure truth and "the sweet voice of the inner Comforter."[25] However, it is not accurate to say that Bernard's use of sensory language is "synaesthetic" in general. Although he uses rhetorical synaesthesia at times (I suggest that he does so when he has in mind the incomprehensibility of how the blessed know God in heaven), such passages should not be taken as all-explanatory or be allowed to obscure other patterns in the way Bernard uses sensory language to articulate theological concepts about God.

In a number of places, Bernard explicitly claims that one sense is prior in the approach to God; that is, more spiritual and more anagogic than another. In the three sermons we just examined, he reproduces the traditional hierarchy that rates sight more "noble" than the other senses (*Div.* 10.4). He also regularly uses the language of sight and light when discussing how a person can encounter God in this life. He not only uses the commonplace metaphors equating sight and thought or understanding but also discusses the kinds of vision in this life (e.g., *SC* 31 and 32), kinds of contemplation and "scrutiny" (e.g., *SC* 62) and illumination (e.g., *SC* 41.3 and 45.5-6). He also maintains the traditional doctrine that the most immediate union that the soul can enjoy is the heavenly *visio Dei*. However, this refers to the perfected sight (or perhaps perfected senses generally) enjoyed by glorified souls after death.[26] It is an immediate relation that requires a purity restricted to angels and those who have "laid aside the body." Those still in the body can only hope to sample it briefly and imperfectly (*SC* 52.5-6).[27] In *Sermon 38* on the Song, Bernard asserts that while the bride has great knowledge of her Bridegroom and her God because she has enjoyed his "frequent colloquies and kisses," she is not ready for a vision of his inaccessible brightness while still living on earth (*SC* 38.3).

Although Bernard uses language of sight to discuss the person's intimate communion with God in this life, he cannot be said to favor it. In a number of places he says that for fallen humanity the primary spiritual sense is hearing (in this he agrees with Augustine).[28] For instance, in *Sermon 53* on the Song, Bernard expands on this order of senses that he finds in Song 2:8:

> Having therefore ascertained the voice, the bride exults with joy saying: 'the voice of my beloved' [Sg 2:8a]. . . . And she continues: 'See, he

> comes leaping over mountains, bounding over the hills [Sg 2:8b].' Having ascertained the presence of the beloved from the sound of his voice, she unhesitatingly turns inquisitive eyes to see him whom she has heard. Hearing leads to sight, 'faith comes from what is heard' [Rom 10:17]. By it hearts are so cleansed that God can be seen; hence you have the expression; 'cleansing hearts by faith' [Acts 15:9]. And thus she sees him coming after she hears his voice; even the Holy Spirit here preserves the order which the prophet described thus: 'Hear O daughter, and see' [Ps 44:11]. (*SC* 53.2; *SBO* 2:96-7)
>
> [Comperta ergo hac voce, sponsa gaudens et exsultans, VOX, inquit, DILECTI MEI. . . . Et addit: ECCE ISTE VENIT SALIENS IN MONTIBUS TRANSILIENS COLLES. Comperta ex auditu vocis dilecti praesentia, incunctanter intendit bene curiosos oculos ad videndum quem audierat. Auditus ducit ad visum: fides ex auditu, qua corde mundantur ut possit videri Deus; sic enim habes: FIDES MUNDANS CORDA. Videt itaque venientem, quem loquentem audierat, observante etiam hic ordinem illum Spirito Sancto, qui apud Prophetam descriptus est ita: AUDI FILI, ET VIDE.]

He goes on to cite Acts 2, where "the coming of the Holy Spirit was perceived first by hearing, afterwards by sight" [Spiritus Sancti adventum primo auditus, dehinc visus percepisse refertur] and mentions that the same can be found elsewhere in scripture (*SC* 52.2; *SBO* 2:97).[29]

In fact, he is often wary about sight as the most "fallen" and deceptive sense—and clearly trading on the connection between sight and (intellectual) knowledge. One of the major structuring motifs of *On Conversion,* for example, is the way fallen spiritual sight can be restored when it hears and listens to the divine voice. Likewise in *Sermon 28* on the Song, where Bernard writes that the soul can only begin to conform to God by listening to his word, in which it finds the truth that is inaccessible to fallen sight (*SC* 28.7). The blackness of the bride in Song 1:4 is a figment of our fallen sight. We can still have access to the truth through the Word of God: "what we hear from him [God] is his, what we see of him is ours" (*SC* 28.3) [quod auditur ex eo; quod in eo videtur, nostrum]. Our sight is fallen and cannot attain to the glory of Christ (*SC* 28.5). The bodily senses cannot possess the truth or wisdom, but deceive the soul. Faith, on the other hand, cannot be deceived, because it is from God and understands invisible truths. Only the hearing that catches the word of God has the truth (*SC* 28.7-8).

THE TOUCH AND TASTE OF UNION, GRACE, AND WISDOM

If fallen sight is healed by hearing, Bernard often articulates the further approach to and union with God using language of taste and touch. Those who acquire faith by hearing can "touch" God in Christ and "taste" wisdom and contemplation. Although God cannot be seen in his unapproachable brightness, "faith reaches the unreachable, catches the unknown, grasps what cannot be measured, takes hold of the uttermost, and in a way encompasses even eternity itself in its broad breast." (*SC* 76.6; *SBO* 2:257-8). [Quid non invenit fides? Attingit inaccessa, deprehendit ignota, comprehendit immensa, apprehendit novissima, ipsam denique aeternitatem suo vastissimo sinu quodammodo circumcludit.] In other words, Bernard inverts the traditional hierarchy: he places sight first and *lowest*, taste and touch last and *highest*. He seems to do so because references to these senses help him articulate the immediacy and mutuality of union with God. In one place, he also uses language of touch to discuss the unity within the Trinity, although he does not develop the idea very far to articulate trinitarian union for us in this life. Bernard's language of taste and touch is primarily incarnational language. Just as he uses strongly tactile language to discuss the conformity of divine and human in Jesus Christ, so he uses it to articulate mystical union between the individual human person and God.

In *Sermon 28* on the Song, Bernard discusses how we attain *unitas spiritus* by faith, which touches what hearing and sight cannot attain. His point is that union is unmediated. True wisdom is obtained not through the physical senses or from human actions, but from God through his word—yet wisdom is something we "grasp" and "touch" when it touches us. We touch Christ when we "anoint" him with thanks in prayer; we thereby touch God, the head of Chris (*SC* 10.8). In turn, "spiritual unguents" flow down from the head of Christ, the divine Son and the head of the church, to touch us through the hem of his cloak—that is, the church and his humanity (*Div.* 25.2). These unguents or "ointments" (Sg 1:2) signify the contrition, devotion, and piety that anoint the bride or soul with virtue and charity and produce faith. Faith is a surer knowledge that "grasps truths unknown to the senses, beyond the range of experience"; by faith, the ascended Christ can "be touched, but by a feeling or movement of the *affectus* (*affectu*), not the hand; by desire, not the eye; by faith, not the [carnal] senses," and then "grasp what is the length and breadth and height and depth" of God" (*SC* 28.9; *SBO* 1:198). [Certiorem fidei cognitionem . . . utique apprehendit quod sensus nescit, experimentum non invenit. . . . Et utique poterit [tangi], sed affectu, non manu; voto, non oculo; fide, non sensibus. . . . Denique [fides] comprehendit suo illo mystico ac profundo sinu, quae sit longitudo, latitudo, sublimitias et profundum.] Christ urges us: "touch me with the hand of faith, the finger of desire, the embrace of love; you will touch me with the mind's eye" (*SC* 28.10; *SBO* 1:199). [Tanges manu fidei, desiderii digito, devotionis amplexu; tanges oculo mentis.] The tactile language indicates the immediate relation of like and like: "Let kind touch kind after the manner of its kind" (*SC* 28.10; *SBO* 1:199; translation from *OSS* 2:97). [Talis talem taliterque tange. . . .] Thus the woman who sought to touch the hem of Jesus' cloak (Jn 20:17) was forbidden to touch it, not because touch itself was unfitting, but because she lacked faith. Her touch was not *like*; her touch was not itself touched and transformed by divine wisdom (*SC* 28.8). The woman was not stirred and transformed *in affectu* such that she could achieve the immediate relation of *true* touch.

Language of touch, then, articulates an immediate, transforming relation. It also implies reciprocity, an idea that helps Bernard emphasize that likeness is restored by faith and that grace gives—or is—*unitas spiritus*. Grace is the touch of the Bridegroom's embrace (*SC* 51.5-6), an "oil of unction" granted by the Holy Spirit (*SC* 14.6, *SC* 8.2). This "anointing" is a "touch" and an "experience" that brings the elect to God through virtue and makes them just such that they taste God's sweetness (*Sent.* 2:23). Grace is food the elect taste and know (*SC* 14.2). Bernard asserts that he longs for this touch and to taste this food, which he knows indirectly, by its odor only (*SC* 14.6).[30] The touch of grace is reciprocal: the elect touched by God's grace can touch God back. They "comprehend" (*comprehendit*) God in his length, width, heigh, and depth. As Leclercq has stressed, "this word [*comprehendere*] taken from St. Paul does not mean 'to know' or 'to understand,' but 'to seize' or 'apprehend'" (Leclercq 1976, 151-52). The elect seize and embrace Christ in union of spirit:

> If you are a saint, you have comprehended and you know; if not, be one and you will know by experience. Holy affect makes a saint, and this affect is two-fold: holy fear of the Lord and holy love. The soul affected perfectly by these comprehends as with two arms, and embraces, binds and holds, and says, 'I held him and I will not let him go' [Sg 3:4].[31] (*Csi.* 5.XIV.30; *SBO* 3:492)
>
> [Si sanctus es, comprehendisti et nosti; si non, esto, et tuo experimento scies. Sanctum facit affectio sancta, et ipsa gemina: timor Domini sanctus et sanctus amor. His perfecte affecta anima, veluti quibusdam duobus brachiis suis comprehendit, amplectitur, stringit, tenet et ait: TENUI EUM NEC DIMITTAM.]
>
> Furthermore he who clings to God is one spirit with him, and wholly changed into divine feeling or movement; nor can he any longer sense or know anything but God and what God senses and knows, for he is full of God. But God is love, and the more one is joined to God, the more full one is of love. (*SC* 26.5; *SBO* 1:173)
>
> [Ceterum qui adhaeret Deo unus spiritus est, et in divinum quemdam totus mutatur affectum; nec potest iam sentire aut sapere nisi Deum, et quod sentit et sapit Deus, plenus Deo. Deus autem caritas est, et quanto quis coniunctior Deo, tanto plenior caritate.]

Note that, in the second passage, Bernard distinguishes "knowing" from "comprehending," seizing, grasping; for "fruition is not in knowledge, but in comprehension" (*Csi.* 5.XIII.27; *SBO* 3:490). [Non in cognitione est fructus, sed in comprehensione.] To seize or to grasp is not to know but to "experience." To be experienced is to be influenced or moved by God, to embrace and hold interiorly. Bernard's point, again. is that the relation established by grace is immediate and reciprocal, just as it a touch.

Note also that Bernard identifies such experience with *affectus*. He often discusses *affectus* using language of touch to emphasize that a relation to God through the *affectus* is immediate, reciprocal, active, and noetic. Affective "wisdom" is tasted, and it is distinct from intellectual knowledge that is "seen."[32] *Affectus* is a complex term; it refers most basically to a transforming influence on the order of grace, and also the human capacity or faculty for that influence. This influence is usually identified with love, and sometimes with other emotions; Bernard usually uses it to refer to our active capacity to desire and love, and our passive capacity to receive love.[33] *Affectus* is the "movement" of the will by which we know God, and sometimes it refers to the will itself.[34] Like the will, the *affectus* can be disturbed and misdirected or rightly ordered by grace. As a faculty of love, *affectus* yields a kind of knowledge (according to Gregory the Great's dictum, *amor ipse notitia est*). Bernard, like many others, pairs *affectus* and affective knowledge with intellect and discursive, rational knowledge, which it complements; but affective knowledge exceeds intellectual knowledge, insofar as God in beyond rational knowing.

Affectus is thus closely associated with union of spirit, which Bernard says is the highest union available to humans in this life. *Unitas spiritus* is a conformity of wills in love; it occurs when the Bride and Bridegroom are joined in one spirit, for "he who adheres to God is one spirit with him" (1 Cor 6:17) (see esp. *SC* 83.2-3). Combining 1 Cor 6:17 with language from the Song of Songs, he associates *unitas spiritus* with touching. The spirit "adheres" to God (*adhaerit*) in an embrace (*amplexus,* Sg 2:6) or a kiss (*osculum,* Sg 1:1a), "holding and not letting go" (Song 3:4).[35] To conform to God in this way is to be wedded to the Word. "Truly it is a spiritual contract, a holy marriage. But more than a contract, it is an embrace. It is certainly an embrace, when identity of will makes one spirit of two" (*SC* 83.3; *SBO* 2:299). [Vere spiritualis sanctique conubii contractus est iste. Parum dixi, contractus: complexus est. Complexus plane, ubi idem velle, et nolle idem, unum facit spiritum de duobus.] He also speaks of "ecstasy of spirit" as an embrace. For Bernard the embrace of Song 2:6 ("His left hand was under my head, his right arm embraced me") must signify intimate contact with the divine. He gives the embrace various interpretations, from ecclesial (i.e., the mutual embrace of Christ and the church [*SC* 79.4-5]) to more experiential and eschatological (the embrace as threat of future punishment and the promise of the kingdom [*SC* 51.8-10; *Dil.* 11.33, *Csi.* 5.XIV.30]; or the action of love or affect by which one experiences the fear and love that bring one back to God [*Ep.* 18.2, *Ep.* 108.4, *Sent.* 3:92]). Bernard identifies the embrace mentioned n Song 2:6 with the image of John reclining on Jesus' breast (Jn 16:25), which is a topos of mystical union, and also with contemplation in which the soul "has retired to solitude and there . . . sweetly sleeps in the arms of the Bridegroom, that is in ecstasy of spirit." (*SC* 52.5-6; *SBO* 2:93; see also *SC* 51.5) [in solitudinem hanc secessisse sponsam, ibique . . . loci inter amplexus sponsi suaviter obdormisse, id est in spiritu excessise.]

Bernard uses the tactile language of "adherence" and "embrace" and "kissing" to articulate the immediacy of union partly because it alludes to the immediate union of God and humanity in Jesus Christ.[36] Unity of spirit is a touch, just as the Incarnation was a touch—a kiss. "Kiss" is the key tactile term used by Bernard to discuss this union. In *Sermon 2* on the Song, Bernard interprets Song 1:1a ("osculetur me osculo oris sui") to refer to the Incarnation: the Word of God or "mouth" assumed human corporeality when it "kissed" the flesh it took on. Just as a kiss is shared by both giver and receiver, so Jesus Christ was both divine and human. The kiss then represents the Christ the mediator, the Word and the incarnate Son of God. We have already heard Bernard refer to Christ's immediate presence as a "kiss." (*QH* 15.2; see above, p. 54). In *Sermon 2* on the Song, after outlining an incarnational interpretation of this kiss, Bernard expands his idea to show how contemplatives can come to God through Christ. The phrase "kiss of his mouth" suggests to Bernard that there are other kisses that came before (*SC* 3). The kiss of the mouth is the most perfect of three: it is a delicious kiss of Christ's lips that can only be achieved after two preceding "kisses of the feet and hands." These special touches of lip and lip outline the life of virtue that leads a contemplative to God. A sinner begins with humility before God, which Bernard likens to "the kiss of the feet." Then he may raise his face from the dust to "the kiss of the hands;" that is, the contemplative further conforms himself to God by temperance and repentance. Only then is he worthy of the kiss of the mouth, which is oneness with divine spirit.

Bernard explains that this anthropomorphic language signifies how God manifests himself in Christ and thereby affects fallen human beings (*SC* 4). We are humble as if at his feet, feel devotion and refreshment as if lifted up at his hand, and rapture as if at his kiss. He writes that the kiss of the feet signifies the saving incarnation of Christ. Trapped in body and oblivious to God's incorporeal action in the world, humanity could not be saved unless God took on a body. So God sent his son to manifest God's power, mercy, and judgement and lead corporeal humanity to the divine spirit. People who acknowledge the saving action of Christ can then be led on to the kiss of the hands and mouth, by which the person feels the love of God that they experience in prayer and in reading the Psalms (*SC* 6-7).

In *Sermon 8* on the Song, Bernard interprets the "kiss of the kiss" ("osculetur osculo oris") to imply the action of God as Trinity. The kiss is the Holy Spirit, who is the love shared by the Father and the Son within the divinity. This kiss is a joining, an embrace, and *touching* that is reserved to the three persons of God. A few fortunate people, such as Paul, can enjoy "the kiss of the kiss," when they are kissed by the Holy Spirit, and participate in the trinitarian kiss and embrace of Father and Son (*SC* 8.8). Bernard does expand very far on this idea. However, his friend William of St. Thierry did; and, partly influenced by William, Hadewijch developed her rich language of touch precisely to articulate the relation within the Trinity and between the Trinity and the human person.

As often as Bernard refers to the sense of touch to discuss union with God, he refers to taste. I have already cited some examples: Bernard writes that theological concepts that express our contact with God "taste," "taste of," and "are tasted." He uses the language of taste (and sometimes eating) to articulate the immediate action of the Holy Spirit in us and its transformative effects. When discussing grace, charity, wisdom, and contemplation Bernard frequently and almost predictably breaks into language of taste (or touch, as we have already seen). I have sketched why: these are all terms that indicate immediate union with God. For example, in *De consideratione* he writes that the third and highest form of consideration "tastes" what the first two hope for and smell (*Csi.* 5.IV; *SBO* 3:469). [Ergo quod prima optat, secunda odorat, tertia gustat.] In *Sermon 50* on the Song, he correlates the highest of three affections with tasting (*SC* 50.4). In *On Loving God* he writes that those who attain the third level of love, who love God for God's sake, "by tasting discover and judge how sweet the Lord is" (*Dil.* 9.26; *SBO* 3:140; see also *Dil.* 15.39 and *Ep.* 11.8) [gustando probari quam suavis est Dominus]. They "taste and see that the Lord is sweet" in charity, contemplation, and wisdom. To love is to taste, and charity grants a taste of God (*Sent.* 2:23, 2:19, 2:23). The taste of God received by the bride when Christ visits in contemplation is retained in the "palate of the heart" (*Div.* 3:1) and draws the *affectus* (*Sent.* 3:97). So the bride enjoys a taste of contemplation (*gustus contemplationis*): "Hence she says: 'And his fruit is sweet to my taste,' signifying the taste of contemplation she received when sweetly raised up by love" (*SC* 48.8; *SBO* 2:72; see also *SC* 67.6). [Unde ait: ET FRUCTUS EIUS DULCIS GUTTURI MEO, gustum contemplationis eius significans, quem obtinuerat per amorem suaviter sublevata.] The taste of contemplation is reserved for the just (*SC* 67.6), who love God and neighbor with perfect charity (*Sent.* 2:19, 2:23, 3:2). Consolation is a taste of wisdom that spurs desire for and love of God (*Sermones in festivitate omnium sanctorum* 1:10). The charity taught by the Holy Spirit grants the soul a "taste of knowledge and a flavoring of grace" (*SC* 8.6; *SBO* 1:39) [et scientiae gustum, et gratiae condimentum]. True, in each of these instances Bernard has in mind that the taste of God is a sampling, a *fore*taste of eternal life and contemplation in heaven. But he does not let that idea crowd out the sensorial reference. *Praelibatio* is the highest attainment to God for those still in the body exactly because it is experiential and immediate, and Bernard plays on the reference to the sense of taste to make that point.

Psalm 33:9 ("gustate et videte quoniam suavis est Deus") is one of Bernard's favorite verses. He cites it to emphasize the affective and experiential immediacy of contemplative contact with God, an immediacy that both originates and issues in true charity and love of God and neighbor. It is notable that Bernard treats it as a verse less about seeing than tasting. He usually emphasizes the verb *gustate,* often omitting the verb *videre* (perhaps because of the echo in 1 Pet 2:3: "gustastis quoniam dulcis est Deus") and sometimes replacing *videre* with another verb.[37] In at least one place, Ps 33:9 suggests a sensory hierarchy to Bernard: "Unless you have tasted, you will not see" (*Conv.* XIII.25; *SBO* 4:99) [nisi gustaveris, non videbis]. This again suggests that sight is the reward of the elect in heaven, taste the highest aspiration of those in the body. The fullness of midday light may be the ultimate goal, but it is inaccessible. "There [in heaven] will be fulfillment; here there is a taste. Therefore taste and see that the Lord is sweet. . . ." (*Div.* 41.12; *SBO* 6,1:253). [Illic erit adimpletio, hic est gustus. Gustamus enim hic et videmus quoniam suavis est Dominus. . . .] The non-elect can have a taste—both a small sample and a kind of contact. Although by this Bernard seems to mean something less than the fullness the elect will enjoy in heaven, it is a gift of sweetness that stimulates the soul to praise.[38] God grants the immediate contact of taste that transforms and leads a contemplative to a purer love (*Dil.* 9.26).[39]

The transforming taste of grace is and imparts wisdom—which, as we have seen, exceeds hearing and sight. It is an "interior taste" that gives a foretaste of divine things (*Sent.* 3:126). Bernard frequently puns on the two meanings of the word *sapere,* "to taste" and "to know," and its etymological kinship to *sapientia,* in order to teach that the wisdom that exalts us to *unitas spiritus* is an immediate, transforming contact with God. We know the God who is spirit by wisdom, not knowledge, and that this wisdom is what we taste in the wine of contemplation.[40] A particularly striking example of this is *Sermon 50* on the Song, where Bernard introduces the term "sapere" at the start and plays with it throughout, deliberately punning on the meanings "taste" and "know" in order to link wisdom, experience, *affectus,* and the highest form of love (e.g., *SC* 50:1). He distinguishes taste as more immediate than olfaction and sight, and he correlates it with the highest, "contemplative" form of love, the "affective" charity that we feel in the presence of God.

This sermon, along with *Sermon 49,* deals with the ordering of love in the soul (Song 2:4, "ordinavit in me caritatem"). *Sermon 50* discusses active and contemplative love. Bernard distinguishes three forms of affection [*affectio*]: that born of the flesh, that ruled by reason, and that seasoned by wisdom (*SC* 50.4; *SBO* 2:80). [Est affectio quam caro gignit, et est quam ratio regit, et est quam condit sapientia.] He alters Ps 33:9 to emphasize that this highest form of love is not an intellectual knowing but an immediate and transforming taste that is the act of knowing in wisdom:

"tertia ab utraque distat, quae et gustat, et sapit quoniam suavis est Dominus. . . ." (*SC* 50.4; *SBO* 2:80). It is "rich and sweet," delighting and refreshing the mind with the sweetness of the Lord (*SC* 50.5). It is a love that is tasted ("sapida caritas") and "the wisdom [*sapientia*] by which all things are tasted and known [*sapiunt*] as they are" (*SC* 50.6; *SBO* 2:81) [sapientia, per quam utique quaeque res sapiunt prout sunt]. God is not tasted and known (*sapit*) as he is, of course, but only according to our power to taste and know ("prout tuum sapere est"); but we can taste and know ourselves when we know ourselves unworthy of any love, except for God's sake (*SC* 50.6). Affective charity sets love in proper, hierarchical order. A person gifted with this charity loves God before all else; as we have heard already, he discriminates between things used and enjoyed "intimo quodam mentis sapore"; and he is wise because he tastes things as they really are (*SC* 50.8).

In *Sermon 85* on the Song, Bernard spins an even more intricate (and untranslatable) net of puns on *sapere, sapientia,* and related words to make the same basic point: wisdom transforms and teaches, and it allows a person to achieve *unitas spiritus* (*SC* 85.8-9). Language of taste helps him articulate these ideas. He again equates taste with ordered affection (*SC* 85.9); but he also makes the link between *sapor* and *sapientia* explicit, emphasizing that wisdom is a taste for goodness in the "palate of the heart" (*SC* 85.8). To be wise is to taste and see that the Lord is sweet; it is to experience (*SC* 85.9). Experienced wisdom is not seen or heard but tasted and touched, because it is an immediate relation to God that is beyond intellectual knowing:

> The Spirit alone reveals it [1 Cor 2:10]: you will consult books to no purpose; you need experience instead. It is wisdom, and man does not know its price. It is drawn from hidden places, and this sweetness is found in the land of those who live sweetly [Jb 28:12-13]. Of course the Lord is sweetness, and unless you taste you will not see. For it is said: 'Taste and see that the Lord is sweet [Ps 33:9]. This is hidden manna, it is the new name which no one knows except him who receives it. Not learning but anointing teaches it; not knowledge but conscience grasps it.[41] (*Conv.* XIII.25; SBO 4:99)
>
> [Solus est Spiritus qui revelat: sine causa paginam consulis; experimentum magis require. Sapientia est, cuius pretium nescit homo. De occultis trahitur, nec in terra suaviter viventium invenitur ista suavitas. Nimirium suavitas Dominus est: nisi gustaveris, non videbis. GUSTATE, inquit, ET VIDETE QUONIAM SUAVIS EST DOMINUS. Manna absconditum est, nomen novum est, quod nemo scit nisi qui accipit. Non illud eruditio, sed unctio docet, nec scientia, sed conscientia comprehendit.]

It seems to be the immediacy of "tasting" that guarantees for Bernard that contemplation is beyond intellectual knowing.

Bernard writes that the taste of God induces a "taste for" divine things: "Having tasted the spirit, one necessarily has no taste for the flesh; having been affected by heavenly things, worldly things are distasteful; having gazed upon eternal things, transitory things are loathsome." (*Ep.* 111:3; *SBO* 7:285). [Gustato spiritu, necesse est desipere carnem; affectanti caelestia, terrena non sapiunt; aeternis inhianti, fastidio sunt transitoria.] This "taste for" is the fruit of grace, and manifests itself in a life of virtue. In *Sermon 27* on the Song, Bernard writes that the bride can fittingly compare herself to the curtains of Solomon (Sg. 1:4)—that is, the attributes of God in all their heavenly beauty—because her life of virtue bears a likeness to heaven: "she seeks and savors the things that are among them [the angels], not the things that are on earth [Col 3:2]." (*SC* 27.6; *SBO* 1:185). [Quaerit postremo et sapit quae apud illos sunt, non quae super terram.] What we seek, we have a taste for and, therefore, a likeness to. Yet we do not acquire what we seek, it is given. Bernard observes that God varies his taste to the "palate of the heart" according to the person (*SC* 31.7; again, "the heart" [*cor*] blends almost imperceptibly into the *affectus*.) The "taste of" and the resulting "taste for" God stokes desire—and confirms the goats among the sheep: "Indeed you had not yet tasted, and you did not know what Christ tastes of, therefore you do not desire; and if you tasted and did not taste and know [*sapuit*] sweetly, you do not have a healthy palate." (*Ep.* 2; *SBO* 7:20). [Aut certe nondum gustasti et nescis quid sapit Christus, ideo que non appetis inexpertum, aut, si gustasti et dulce non sapuit, sanum non habes palatum.] (Conversely, Christ tastes our virtues [*SC* 71.14, 72.3] and distinguishes the damned from the elect by means of taste [*Sent.* 3:22].)

Perhaps it is unsurprising or even inevitable that Bernard refers to tasting and eating (and food, e.g., "manna," above) as metaphors for knowledge and learning, because the association is apparently ubiquitous, and it is certainly common in the Bible and early Christian writings. Origen is particularly fond of using them with the exegetical meaning that we taste true doctrine in the scriptures, the "living bread" and the "hidden manna" that lurks within the "letter that kills" (e.g., *SC* 1).[42] Bernard, too, writes that the words of Scripture are the bread of life in which we taste truth, wisdom and Christ the Word, who nourishes the soul and tastes sweet (*SC* 1, 67.1, 53.5). If we taste scripture and doctrine, what we gain is a foretaste in contemplation of the God beyond knowing (*Sent.* 3.96). The notion of eating overlaps with that of tasting, both conceptually and in Bernard's usage. On the other hand, Bernard recognizes that references to eating imply engulfing and assimilation, not merely immediate contact (a point I would also make, extending the phenomenological account of sensation I have borrowed from Hans Jonas).[43] Bernard uses language of both taste and touch to articulate the immediate relation between the person and

God. However, he feels he must qualify his language about mutual eating in order not to imply union without difference. In *Sermon 71* on the Song, he describes union as a conformity in which the Bridegroom eats the soul and transforms it: "I [the bride] am chewed as I am reproved by him; I am swallowed as I am taught; I am digested as I am changed; I am assimilated as I am transformed; I am made one as I am conformed" (*SC* 71.5; *SBO* 2:217). [Mandor cum arguor, glutior cum instituor, decoquor cum immutor, digeror cum transformor, unior cum conformor]. (Note that Bernard implicitly distinguishes eating from tasting, which is conspicuously absent from the list.) But note further that what is transformed is not the eater, Christ, but that which is eaten, the soul. This complements the notion that as we "eat" doctrine we are nourished. However, Bernard is uncomfortable with the idea of transformation in this passage, and later in the same sermon he qualifies it. He specifies that the mutual eating of the bride and bridegroom signifies not total assimilation and transformation into one substance or nature but a more discrete engulfing or containment. Like the Father and the Son, the bride and the Bridegroom "are capable of containing each other without being divisible, and of being contained without being divided" (*SC* 71.7; *SBO* 2:218-19) [. . . sane ita capabiles ut non partibilies, ita capaces ut not participes].[44] He stresses not assimilation—chewing, swallowing, digesting—but a particularly intimate *amplexus*, a mutual *touching*.

Bernard uses the language of taste, like language of touch, to articulate the anagogic character of experience, wisdom, and ordered love. He seems to do so because the immediacy of taste helps him articulate the immediacy of our closest conformity to God. His usage is significant to his ideas about mystical union because it suggests that this union is immediate, which is somewhat at odds with his explicit statements about *unitas spiritus*. It is also at odds with the dualist elements in his concept of the human person, because it implies that a person can know God by means of a single sensorium that belongs to the whole person. He seems torn between the idea that we are essentially souls, exiled in the body and estranged from God and heaven, and the idea that we are integrated persons, equally soul and body, who will be united to God in the body and who can achieve an immediate relation to God now, in this life, in our "experienced" bodies, in the pattern of the God-man Jesus Christ. Despite these ambiguities, Bernard's concept of spiritual sensation is less dualist than the concept of the spiritual senses developed by Origen, Augustine, and most of the scholastics who wrote about it after Bernard.

CHAPTER 4

Hadewijch:

The Touch and Taste of *Minne*

Bernard uses language of taste and touch to suggest that *unitas spiritus* is an immediate relation, but this idea rubs along beside his careful statements that *unitas spiritus* is a union of wills, not essences, and that the person always remains subordinate to and distinct from God. Hadewijch of Brabant (fl. 1220-1250?)[1] is more radical. She fully exploits the intimate bodily relations implied by language of taste and touch to articulate her idea that union with God is an immediate, reciprocal, and dynamic union of essences. Her ideas about the senses of taste and touch and about the relations that attach to language that refers to those senses match her theological ideas about the nature of God and the structure of the human person. She assumes that to sense by taste and touch is to make immediate and continuing contact with another bodies and thereby acquire knowledge that alters our judgements and desires. She also assumes that to taste or to touch is by definition to *be* touched: it is reciprocal. Touching, in her mind, also has another characteristic: to touch is also to stir up, to move violently and change.

Hadewijch also presents union with God as an immediate, reciprocal, and turbulent exchange. That is how she understands the relations between the persons of the Trinity, and she conceives of union as participation in these relations, or even an aspect of them. The bodily reference of tactile and gustatory language also agrees with her ideal of holy life and the concept of the human person it assumes. Even more than Bernard, Hadewijch observes no strong distinction between matter and spirit, soul and body; she assumes that people know, become like, and can be unified with God as an integrated whole, soul and body. Also like Bernard, she models this intimate connection on the incarnation of God in Christ; but she develops the model somewhat differently. First, she gives more emphasis to Christ's bodily presence in the Eucharist. Several times associates union with God and partaking of the Eucharist, when we taste and touch Christ both spiritually and physically.[2] Bernard, in contrast, says little about the Eucharist, and nothing about how it relates to union. Second and more significantly, Hadewijch assumes that a person cannot become one with God unless they do and act as the human Christ did in the world. In accordance with the thirteenth-century ideal of apostolic service in the world, Hadewijch emphasizes that we become like Christ only by following him in a life of physical, suffering service in the world. If a person serves and loves their neighbors, despite scorn and rejection, they share in the actions of both the human Jesus and the divine Christ and, according to Hadewijch, participate in the dynamic relations of the Trinity itself, attaining Christ-like spiritual status and achieving union with God. True, Hadewijch insists that union springs from the will: when a person wills as Christ willed, they do as he did, and they are as he was (and is). But she also teaches that such perfect willing *necessarily* entails physical, embodied action. Perfect willing and physical action in the world are not separable, or even distinguishable, because to will and to do as the human Christ is to will and do as God—or, as Hadewijch conceives it, to will and do as *minne,* "Love." Identification with *minne* allows a person to achieve ontological union with the ground of the Trinity and the root of God.

Hadewijch developed her theological concept of *minne* on the basis of ideas from the literature of courtly love. (I will usually use the Middle Dutch word *minne* because the English word "love" captures only some of its meanings and suggests others that are misleading.) The concept *minne* is complex but absolutely central to Hadewijch's ideas about God, and understanding it is central to understanding why she uses language of taste and touch to discuss the nature of God and union with God. I will discuss it in some detail a little later on. For now, the key thing to know is that Hadewijch thinks of *minne* as more than an emotion; it is equally an ontological principle that binds the Trinity to itself and to the human person—an essential something that is simultaneously fully divine and fully accessible to embodied people. *Minne* is something that is equally God, the person, and the bond between the two. Shared *minne*-nature allows a person to achieve a relation with God that is immediate and fully reciprocal. Hadewijch also presents *minne* as intrinsically dynamic and turbulent. Therefore, she presents the bond of *minne* the same way: in union, *minne*'s dynamism touches a person and stirs them to a restlessness that is like itself and like the internal life of the Trinity. More than merely "like," the restlessness of *minne* is itself that bond: in this earthly, embodied life, the bond of *minne* is felt and known not as tongue-tying emotional exaltation or contemplative peace but as palpable turbulence that, as in the Trinity, pours forth words and works. It is a hard, unyielding cord of suffering that we know not as spiritualized, disembodied souls but in body and in the world, as whole people, as Christ did.

In this way, Hadewijch presents union as both trinitarian and christological; in the human person, it is a fact of both will and physical body. We have seen that Bernard worked with a largely traditional monastic anthropology, tightly bound to christology and based on language of image and likeness (*imago et similitudo*) that has little to do with the Trinity. Hadewijch's theology, on the other hand, combines the theme of image and likeness with the idea that human beings have emanated, flowed, or been born from the Trinity—more precisely, from *minne,* which is the inner dynamism of the Trinity. Those "born from *minne*" participate in divinity ontologically and so profoundly that it is possible for them to exist within the inner dynamism of God and achieve a unity without difference. Because of and by means of *minne,* a person can gain divine status and knowledge and can attain an immediate and fully reciprocal relation with God. Language of taste and touch helps Hadewijch develop these ideas because it implies a relation of bodies that is immediate, reciprocal, and dynamic. It therefore helps her articulate her main teachings: that union with God can be achieves only by way of love, *minne,* and that to achieve this union is simultaneously to achieve likeness to the human Christ, the God who perfectly served *minne,* who suffered, and who felt and knew through the body.[3]

In one respect, Hadewijch's idea that we taste and touch God by *minne* echoes Origen's Platonic dualism. I argued that Origen assimilates the spiritual senses to mind or spirit, then correlates spirit "here" to spirit "there." Hadewijch, similarly, treats *minne* as the nonbodily something that is like God and so allows us to "sense" him. Yet Hadewijch in no way denies or even minimizes the idea that *minne,* like language of taste and touch, refers to the sensing body. Rather, she insists that, by imitating Christ (as by ingesting the bread of the Eucharist) we know and achieve unity with God as whole persons and with the help of our bodily senses. We do the same by living the life of *minne.* Its bodily and sensory aspects in the world cannot be segregated from its ontological aspects, which name the deepest nature of God and the human person. In order to talk about God, Hadewijch appeals both to the embodied self and to the spiritual aspect of *minne.* Indeed, she refers to the embodied self *in order* to discuss the spiritual aspects of *minne* and God. She assumes at every turn that the human person is not a spiritual essence trapped in matter but one single soul and body. This integral anthropology is evident both in her theological teachings and in the language she uses to develop them.

GENRE, HADEWIJCH'S NARRATOR, AND THEOLOGICAL CONCEPTS

It is easy to miss or misinterpret these ideas because Hadewijch's choices of genre and style of narrative lead modern readers to interpret her writings as accounts of her "experience" and overlook or ignore the theological concepts that she teaches. Although Hadewijch wrote in established genres for religious writing—didactic letters, visionary narrative, and poems on divine love (less traditional, although they had some antecedents)[4]—she did not write biblical commentary or formal theological treatises, and she wrote in a dialect of Middle Dutch. That is, she did not write in the genres or language of monastic or scholastic theological disputation. She and her audience were neither monks nor schoolmen. Although we know next to nothing about Hadewijch's life and situation, she was probably a beguine—that is, not a regular nun, but an unmarried lay woman living a quasi monastic life in a town: praying, reading and doing public service, such as caring for the sick and teaching girls, and living either with her family or with similar-minded women in a shared house.[5] The beguine "movement," if we can call it that, was an amorphous phenomenon, but its ideals are clearly close kin to those that led Francis of Assisi and throngs of other men to take up the life of apostolic poverty and service in the world. Hadewijch and her audience had some education and may well have known some Latin, but their religious life was spoken in the vernacular and, as women, they would have had at best limited access to most theological writings and the text of the Bible itself (although this did not keep them from having large chunks of it by heart).[6] They were not allowed the university education that would make formal disputation on set topics possible (and appealing). In any case, women (like all lay people, but more so) were not allowed to dispute or preach on theological doctrine, and were forbidden to interpret and comment on the Bible.[7]

However, that does not mean that women were uneducated, did not write theology, or lacked significant theological ideas. Hadewijch was perhaps exceptional, but she was quite well educated and seems to have read a fair range of theological writings.[8] She took up their ideas and developed them in her own way. Two features of her writings aspects are especially significant for my argument. First, Hadewijch does not cite the Bible that much, and she only rarely develops her ideas by interpreting a Bible passage. Second, she (like some other women religious writers in her day [McGinn, *Flowering,* 12-30]) writes about a distinctive narrative persona: a woman struggling to live a certain kind of holy life in the world, who seeks and achieves the very highest spiritual status with respect to God and must deal with the consequences. Often she writes about this persona in the first person, sometimes in the third. The absence of a biblical context and the presence of a strong narrative persona influence both what her theological ideas are—including her ideas about sensory language and its relation to union with God—and how those ideas are interpreted, by both medieval and modern audiences.

In these respects Bernard's writings are a useful contrast—as are, for that matter, the writings of Origen and the other theologians I have discussed so far, but I will focus on Bernard, because he and Hadewijch use sensory language in a similar way and have similar assumptions about the spiritual senses. Bernard, like most theologians in his day and before, developed his ideas primarily by biblical interpretation. Even his works that are not commentaries are thick with biblical echoes, allusions, and references. I have argued, more specifically, that anchoring language of taste and touch to biblical usage helped Bernard constrain its inadmissable somatic reference and maintain its spiritual meaning, both for himself and for his audience of Cistercian monks. More subtly, Bernard anchors this spiritual meaning by associating embodied experience very closely with exegesis, especially exegesis of the Song of Songs.

Hadewijch does neither of these things. First, she does not write biblical commentary and does not even cite the Bible very often. Joris Reynaert has shown that, in both her prose and her poetry, identifiable citations are less common than some modern editors would have us believe.[9] They are certainly much less common than they are in writings by male monastics, such as Bernard of Clairvaux. She cites key verses to support her points and very occasionally offers more sustained comment on a short biblical passage. Of those biblical references that can be identified with confidence, the overwhelming majority are brief phrases.[10] In only five instances does she comment on biblical passages by rewriting or paraphrasing them. As a result, although Hadewijch sows her writings with an armful of authoritative biblical phrases and ideas (especially certain verses from Job, John, and Matthew), she was no biblical exegete. Of course, if she did know the Bible well or had access to some of it in written form, it would have been dangerous for her to write about it in detail or at length; the Bible was the preserve of priests and monks. Whatever her reasons, Hadewijch did not articulate her ideas in biblical language, with biblical motifs, or in the context of biblical exegesis. One consequence is that the language of taste and touch she uses lacks an halo of biblical authority and a clearly-defined spiritual meaning that her audience (then and now) would recognize.

Second, unlike Bernard, Hadewijch does not help give somatic language a spiritual meaning by binding her protagonist and her reader together in the act of biblical interpretation; in particular, she does not use the Song to spiritualize, simultaneously, both highly somatic sensory language and the language of love. (Here the absence of a biblical context in Hadewijch's writings intertwines with the characteristics of her narrative persona; in a moment I will discuss her narrative persona independently of these biblical issues.) Bernard assimilates his protagonist, the lover who seeks God, to a spiritual ideal defined by an authoritative spiritual text and its corona of spiritual exegesis, and this helps him fix a safely circumscribed, authoritative spiritual meaning for the sensory language he predicates of that protagonist. Like Origen (in fact, influenced by Origen), Bernard treats the text of the Song as the text of the self. He interprets one of the speakers in the Song, the female "bride," not just as the church collectively wedded to

Christ, but also and more profoundly as the individual monk (at least insofar as that monk desires and seeks God in love). In his sermons he challenges his monastic audience to find themselves in this biblical protagonist, and so find an intimate relation to God that is still mediated by the Bible. As result, when Bernard appropriates sensory language from the Bible to talk about this immediate relation, that language refers *both* to the self *and* to the Bible. It is thus is both textual and "experiential," both spiritual and corporeal.

Hadewijch also deploys language of taste and touch within an exposition of divine love, but neither the language nor the exposition have much to do with the Song of Songs. Although she probably had some knowledge of monastic Song exegesis, at least at second hand (she certainly had heard the Song preached on: see *Visions* 9:4-6), Hadewijch does not develop its characteristic "bridal" themes.[11] Rather, as Barbara Newman has emphasized, Hadewijch develops her concept of love from the conventions and motifs of courtly love or *fin amour*. She presents her protagonist, the "self" who loves God, not as the "bridal self" developed by generations of monastic interpreters of the Song, but as the "courtly self," the lover we find in vernacular love poetry (the masculine *fin amant* and its matching variant, the feminine *amie*). As Newman draws the contrast, the monastic "bridal self" is fearful because of sin but still hopeful and confident that it has a stable relationship to God. The "courtly self" that Hadewijch develops, on the other hand, joys and suffers within a relationship that is essentially unstable and hopeless.[12] Of course, the poetry of *fine amour* was itself influenced by the Song and its monastic interpretations, so they are not absolutely isolated from one another. And, of course, the "courtly self" and its model of love is no less and ideal than the "bridal self" of the Song exegetes; but its somatic and frankly carnal significance was stronger and more overt, less constricted and suppressed. In this respect, Hadewijch's protagonist is detached from and independent of the lover(s) in the Song of Songs and lacks the explicitly and authoritatively spiritual and theological significance granted to that lover by a long tradition of biblical interpreters. Hadewijch does not present her narrator as an ideal "I" whose human physicality is a fiction that serves discourse about nonmaterial, spiritual things but, rather, as an ideal "I" in a suffering, flesh-and-blood struggle to love and achieve union with God.

Many modern readers would say there is a simpler way to put this: Hadewijch writes about herself. That is, the narrator in Hadewijch's writings is Hadewijch herself, the author, a physical woman who lived a certain kind of holy life. It is common to interpret Hadewijch as an example of the thirteenth-century women (and some men) who wrote about God by writing about themselves and their own experience—their actions, thoughts, and feelings. However, no narrative "self" is identical with the author. Medieval people knew this perfectly well.[13] So do modern scholars, in principle. In practice, however, especially when interpreting the religious writings of medieval women, some scholars tend to forget or minimize the distinction between the author and the literary persona, at least partly because they are eager to mine these writings for autobiographical testaments to medieval women's "experience" (of the world and/or of God). Scholars commonly seize on sensory language, in particular, as a transcription of "experience" that gives us evidence of God's action.[14] As I discussed in chapter 1, I think this type of interpretation has serious problems. First among them is that it overlooks the intellectual and formally *conceptual* contributions of medieval women to theology in its abstract, speculative aspect. An "experiential" interpretation is especially inappropriate for Hadewijch's writings because they are not especially autobiographical. Hadewijch does not give us a self-portrait. To the extent that she writes about the actions, thoughts, and feelings of her protagonist(s?), she does so to articulate her ideas about God, union with God, and how to achieve that union.

Some modern scholars have paid significant attention to sensory language in Hadewijch's writings, but all have given it an "experiential" interpretation and so largely overlooked how Hadewijch uses it to articulate concepts about God and union with God. John Giles Milhaven has drawn attention to the way Hadewijch uses strongly somatic language of taste and especially touch: embracing, kissing, and eating (Milhaven, 101-10). He interprets this usage as an artifact of her "bodily knowing." He believes that Hadewijch, in a way characteristic of women, uses somatic language to "describe" the experience(s) by which she loved and knew God (Milhaven, x).[15] He claims that Hadewijch, like other religious women of her era, would not have considered her descriptions mere "metaphors" for spiritual knowing in which her body played no part, but as direct descriptions of what happened. He also argues that she knew such knowing was not "ordinary sensory experience" (Milhaven 103, 88). However, if Hadewijch's "experience" was not "ordinary sensory experience," and she uses sensory language to describe that experience, then the sensory language she uses *must* be metaphorical. Milhaven wants it not to be, but he cannot escape the dilemma because he will not let go of his assumption that sensory language transcribes experience (Milhaven, esp. 84-89, 101-3, and 15-16).[16] Because he wants to unearth evidence for his idea of "women's experience" from Hadewijch's writings, Milhaven does not address how Hadewijch's use of somatic language is governed by and in turn articulates her more speculative theological ideas about the human person, the Trinity, and their union in *minne*.

Neither does Caroline Walker Bynum (Bynum, 1988). Although she only touches on Hadewijch's use of sensory language, her interpretation demands careful attention because it has been widely influential. Bynum's focus is mainly references to eating and drinking; as I discussed in the last chapter, I consider these distinct from tasting because they imply different relations between bodies, but Bynum does not make this distinction and cites medieval use of sensory language to support her points (Bynum 1988, 153-60). Bynum tends to interpret language of eating as bodily language that reports or testifies to experience. She argues, for example, that Hadewijch uses language of eating and drinking to talk about intimate union with God because she, like other medieval women Bynum studies, espoused a style of piety that emphasized we can know God in the body and *with* the body. She argues that, for Hadewijch, "*to eat* and *to be eaten* express that interpenetration and mutual engulfing, that fusion of fleshly humanness with fleshly humanness, that Hadewijch saw as necessary for uniting with a God-who-is-man" (Bynum 1988, 156). Metaphorically, these verbs make sense because they have eucharistic "overtones of physicality joining physicality" (Bynum 1988, 154) that correlate with and refer to physical acts or practice that were considered pious: especially the punishing feats of feeding, eating, and fasting that Bynum collects from the *vitae* of medieval holy women.

However, it is not at all clear that Hadewijch was an exponent of this style of piety. She tells us very little about ascetic practices of her own, and although she extols *imitatio Christi,* her main concern is not the physical details of such action but the psychological and ontological state of "living in God's will."[17] Hadewijch uses strongly somatic language, but not to discuss a performed piety centered on physical acts. Bynum presents Hadewijch as an exponent of such piety and emphasizes an "experiential" interpretation of Hadewijch's use of somatic language; this is in accordance with her assumption that medieval accounts of holy women give us evidence of those women's religious and mystical experience:

> The tendency of women to somatize religious experience and give positive significance to bodily occurrences is related to what is generally recognized to be a more experiential quality in their mystical writing. Male writers too, of course, use extremely physical and physiological language to speak of encounter with God. . . . But men's writing often lacks the immediacy of women's; the male voice is impersonal. . . . men write of "the mystical experience" giving a general description that may be used as a theory or a yardstick, whereas women write of "my mystical experience," speaking directly of something that may have occurred to them alone. . . . Their [women's] descriptions of themselves or of other women often, from a modern point of view, hopelessly blur the line between spiritual or psychological, on the one hand, and bodily or even sexual, on the other. . . . in order to communicate and share a piety in which spiritual-somatic experiences lie at the center. (Bynum 1991, 190-91)

Bynum tends to ignore these women's ideas about more speculative theological topics, such as the Trinity and the nature of the soul, apparently because she thinks of medieval holy women as "doers" rather than "thinkers." For example, she is not very clear about *why* they wrote about God in a way different than men: she seems to say that they wrote about themselves not, like men, to get at concepts ("a theory) or to lay down a model for others ("a yardstick"), but to "communicate and share" their style of piety. She comes very close to treating these writings as transcriptions ("immediate," "direct") of the women's bodily and mental states—their "somatized religious experience" and "spiritual-somatic experiences." In other words, Bynum, like so many others, treats this women as active makers mainly with respect to their experience, not with respect to ideas about salvation, God, and human access to them—more abstract or formal theological ideas are only a shadow to their experience. Bynum does not distinguish between an author who writes about experience (the rhetorical technique) and recording an experience (the stenographer's activity), so she tends to overlook the theological content of the texts she examines. Similarly, she overlooks the fact that most of the surviving accounts of medieval women were written by men. As Amy Hollywood has noted, accounts of holy women written by men very often celebrate those women's strongly somatic and highly visible piety, but those penned by holy women themselves emphasize theological concepts about the will, the nature of God, the structure of the Trinity, and other topics that pertain to their inner lives and ideas. She notes that Bynum does not account for this difference.[18] Hollywood argues that Mechthild of Magdeburg and Marguerite Porete reject the somatic piety attributed to women; I would add that Hadewijch does too.

Bynum's "experiential" interpretation of Hadewijch's writings is not adequate to their theological sophistication. I think we must assume that Hadewijch, like other medieval authors of both sexes, *chose* to use strongly somatic language, just as she chose to write in this or that style and genre, and that she did so less "to communicate and share a piety in which spiritual-somatic experiences lie at the center" than to articulate complex ideas about God. One scholar who has taken this tack is Joris Reynaert. In his study of Hadewijch's imagery (Reynaert 1981), he carefully analyzes how she uses sensory language and reveals a good deal of its conceptual content and theological significance. He classes sensory language with "somatic language" and, unfortunately, assumes that it "records" somatic experience (Reynaert 1981, 207, 189-229). Even though he argues that Hadewijch's use of language is simultaneously a literary, linguistic, and "philosophical" matter, and even though he thinks we can not use her imagery as evidence to make claims about her "psychological temperament," much less the conditions of her life, nevertheless, he judges that Hadewijch's imagery is "perhaps open to psychoanalytic interpretation" because "the meaning [of her imagery] lies . . . on the level of *religious* experience" (Reynaert 198, 49-50). What he means by "psychoanalytic interpretation" is less clear than his basic assumption: if Hadewijch's writings provide us no strong evidence for claims about her worldly experience, they *do* give us good evidence for claims about "religious" and/or "mystical experience." Again, this suggests that Hadewijch wrote less about God than about herself, and the sensory language she uses is data about her.

Another Dutch scholar, Frank Willaert, has made the essential point in his analysis of Hadewijch's book of *Visions:* Hadewijch did not write this book as a transcription of "what happened," addressed, like a diary, to herself; rather, she wrote it primarily as a book of instruction addressed to her friends or followers (Willaert 1986, 368-87).[19] He argues that Hadewijch's audience probably assumed it spoke of experience, and that this probably affected the way she wrote. Willaert's argument suggests (although he does not say this explicitly) that Hadewijch drew on the *idea* and the *rhetoric* of experience in order to teach about God.

As I have already mentioned, I see no reason to think this was either a calculated rhetorical ploy or a thoughtless assumption. Hadewijch herself may well have conceptualized some of what she thought and wrote as "experience." That is, she may have written about "experiences" not just to be understood and persuasive, but because she may well have thought about theological concepts in experiential terms—as narratives of remembered and imagined encounters with God. Like others of her time and place, she probably expected to know God that way, especially since the church taught that, as a lay woman, her only access to God aside from the sacramental life of the church was by means of extra-ordinary personal visitations of God in the Spirit. Although she does not dwell on the subject, Hadewijch, like other devout lay people in her day, probably appealed to experience as one important way to establish her authority to speak and write theology. On the other hand, she and her audience were perfectly familiar with the rhetorical and conceptual complexities of narratives, "experiential" and otherwise. It is misleading to assume that they understood experiential language in any simplistic way (much less that they understood "experience" according to any of its modern, often empiricist, meanings), and it is a mistake to interpret the sensory language in Hadewijch's writings as a record of her sensory experiences.

SENSORY LANGUAGE AND SPIRITUAL SENSES

I argue that Hadewijch, in general, writes in a genre and a style that suggests a physical protagonist actively sensing God with her bodily senses and, in particular, uses language of taste and touch to explain and promote her theological ideas about the structure of the human person and the Trinity and the profound ontological unity between them. In the process, she implies a concept of spiritual sensation based on a single human sensorium. Hadewijch does not explicitly mention the spiritual senses. She makes one passing reference to the "senses of the soul (*sinne zielen*) that await . . . spiritual wonders," which might be the hearing, sight, and taste that sense heavenly things, but might be distinct from them (*Letter* 28:156). However, this is in a letter Hadewijch probably did not write (see note 1, above). In her undisputed writings, Hadewijch only refers more vaguely to the "inner senses" (*inneghen sinnen*). This phrase is highly ambiguous. *Sinne* is hard to render in English but usually means something close to "mind," and so *inneghen sinnen* could easily refer to powers of the mind.[20] This accords with the Latin phrase *sensus interiores,* which in twelfth and thirteenth theological texts usually refers to the powers of the soul or the functions of the mind—very often the Augustinian formula of intellect, memory, and will. (The term *sensus interiores* is, usually, conceptually distinct from the term *sensus spirituales,* the spiritual senses.)

It is not clear if Hadewijch was influenced by the meaning of the Latin phrase; it is perhaps possible that she uses *innegehen sinne* to refer to spiritual senses or sensations (i.e., not necessarily autonomous organs, but distinct actions). She writes that the inner senses have as their object *minne,* divine Love, and her personified actions. She writes that "above all, you [*minne*] are the delight of the inward senses" (*PS* 42:28) [boven al sidi {minne} ghenoechte den inneghen sinne] and that "Judgment of *minne* / goes deep within / by means of the inward senses; . . ." (*PS* 36.47-49) [Vonnesse van minnen / Gheet diepe binnen / Met inneghen sinnen; / Die en mach bekinnen . . .].[21] The inner senses also receive knowledge of God from enlightened reason, which is ultimately incapable of it: "Enlightened reason shows the interior senses a little of God" (*Letter* 22:17). [Verlighte redenene toent den inneghen sinnen een lettel van gode.] Most interestingly, in *Stanzaic Poem 25* she writes about the person's relationship with God using language of hearing, touch, and taste, and then implies that this sensory language refers to some part of the human person, to "inner senses" (*inneghen sinne*) by which it knows *minne:*

> The hidden ways that *minne* sends me
> are those that completely rob me of myself.
> That noise, that high, loud gift
> of low silence makes me deaf.
>
> Her low silence is unheard,
> however high and loud the noise she makes,
> except by she who has experienced it,
> and whom she has completely sought for herself,
> and so closely touched with her deep touch
> that he feels himself completely in *minne.*
>
> As the mind with wonder thus thoroughly tastes,
> the noise ceases for an hour;
> alas! soon desire awakes, and wakes
> the inner senses with a new storm. (*PS* 25:27-40)

> [Die verhoelne weghe die minne mi sent, / Die sijnt die mi van mi al roeven. / Dat gherochte, dat hoghe prosent / Der nederre stillen, doet mi verdoven. // IV. Hare nedere stille es onghehoert / Hoe hoghe gheruchte dat si maect, / En si allene dies hevet becoert / Ende dien minne in hare al hevet ghesaect, / Ende met diepen gherijnne so na gheraect, / Dat hi hem al ghevoele in minne. // Alse sine met wondre also doresmaect, / Cesseert een ure tgheruchte daerinne; / Ay, saen wect begherte die waect / Met nuwen storme die inneghe sinne.]

Although these stanzas seem distinguish between inner and outer senses, they are the only place in Hadewijch's writings . But it is an isolated reference, and does not agree with Hadewijch's main assumptions about theological anthropology.

So if Hadewijch knew about any explicit doctrine of spiritual senses, in Origen's version or its later reworkings, she does not use it. On the other hand, she probably *was* influenced by Bernard. There is some evidence she knew Bernard's works or at least some ideas he made popular.[22] In any case, there is little question that Bernard and Hadewijch worked from related and generally available theological concepts. It is equally clear that they interpret and use those concepts somewhat differently and that, even though they are equally willing to use language of taste and touch to discuss God and how we know God and end up with similarly integral concepts of spiritual sensation, they work with somewhat different theological ideas.

MINNE

I have already briefly mentioned Hadewijch's complex concept of *minne*. We must now examine it further, because it is the dynamic core of her theology, and it governs how she uses language of taste and touch. Hadewijch uses the term *minne* not only to indicate the emotion "love" and the actions that follow from it, but also to plumb the paradoxical and dynamic inner nature of God the Trinity and the identically paradoxical and dynamic nature of the bond between God and human beings.[23] This bond is *minne*, and Hadewijch articulates it using language of taste and, especially, touch. She seems to do so because she assumes the sense of touch establishes a relation between bodies that is immediate, reciprocal, and dynamic; it is therefore uniquely appropriate to the ontological relation between human beings and God, founded on *minne*, which she also conceives of as immediate, fully reciprocal, and dynamic. Her use of this sensory language both derives from and articulates her assumption that those who, like Christ, perfectly serve *minne* thereby know and become one with God the Trinity; and they do so not as souls refined from the world but as whole people living in it.

Minne, in Hadewijch's formulation, is a paradox, a riddle that the reasoning mind cannot fathom.[24] She plainly developed it from the conventions refined by Northern French poets of *fin amour,*[25] which she integrates with theological interpretations of love well established and widely known in the thirteenth century. She reproduces common Augustinian ideas and other concepts that she seems to know from twelfth-century writers, especially Richard of St. Victor and William of St. Thierry, but also Bernard of Clairvaux.[26] Among these concepts are the basic assumption that love is the foundation and goal of our relation to God, and its matching commonplace, God is love; the idea that the inner life and bond of the Trinity is love; and the idea that the divine lover is alternately present and absent.

Hadewijch leaves none of these ideas untouched by the conventions of *fine amour,* in which relations between lovers are based on feudal notions of service and reward, rights and duties, debts and demands. For example, in her cycle of *Stanzaic Poems,* she articulates a complex and subtle relation of presence and absence, enjoyment and suffering, but does not do it in the traditional way, using the Song of Songs and based on the model of courtship and marriage. Rather, she personifies *minne* as a demanding and capricious feudal lady, desired by an endlessly suffering but loyal and courageous lover. The devoted lover of *minne* desires to be with her, serves her faithfully and selflessly, but she responds unpredictably. Sometimes she rewards her lover for his service, and sometimes she disdains, rejects and ignores him. Hadewijch insists both aspects are essential to the nature of *minne.* A lover who only wants rewards is no true lover, or does not love true *minne.*

Hadewijch's personification of *minne* partly emphasizes that a life of holy service to God is difficult and psychologically complex, and partly helps her unfold the essential nature of the conceptual entity *minne*—that is, that aspect of *minne* distinct from the world and distinct from psychological states in the person; what we might call *minne* the cosmic principle. The "blows" rained on the lover by *Frau Minne* are outward manifestations of her essential tempestuousness. *Minne*'s storminess and the "storms" she stirs up in her lovers are essential to *minne,* as are her "sweetness" and the "satisfaction" and "enjoyment" she gives her lover. Both together form her paradoxical, unstable nature. The idea that love is unstable echoes Richard of St. Victor, who in his *Four Degrees of Violent Charity* analyzes love of God as a difficult and in some ways unmanageable psychological force.[27] It contrasts with the more common formulation elaborated by Bernard of Clairvaux, who, although attentive to the psychological difficulty and vagaries of seeking to love God, did not emphasize that love itself was inherently turbulent. Although he speaks of *amor vehemens,* Bernard's point is that the troubles of *amor Dei* result from *our* instability, from the instability of love in *us,* not from the instability of love itself (see, for example, *SC* 74.4). Hadewijch, however, unlike both these men, personifies *minne* not only to dramatize the complexities of the emotion(s) of love, but also because she conceives of it as an autonomous being that has an essential and essentially turbulent nature unto itself.[28] She presents *minne* as more than a power or force that binds human beings to him and allows us to be one with him; she presents it as a shared "something" whose dynamic "nature" (*nature*) or "being" (*wesen*) is the fundamental core—the "ground" (*gront*) or "depth" (*diepheit*)—of both God and human beings.[29] "*Minne* . . . is God himself by nature" (*Letter* 19:31-32). [Minne die god selue bi naturen es.] Hadewijch associates *minne* not only with God as a whole (*Visions* 3; *Letter* 19:27-36) but also specifically with the Son (*PS* 29:41-43) and with the Holy Spirit (*Visions* 2). She seems to do so because she thinks of *minne* as the "oneness" within God, the dynamic binding power that embraces the Trinity such that each of the three divine persons is *minne*.

The relation between *minne* and the Trinity is perhaps most evident in the way Hadewijch characterizes both the internal economy of God and *minne* as "flowing." For example, in *Letter* 22 Hadewijch discusses the paradoxes of God in a typically Neoplatonic schema of *exitus* and *reditus*. She writes that God is the source of all and sustains all, flowing out from the persons of the Trinity to created things; and that likewise, within God, the "depth," "abyss," or the "ground of his eternal nature" is the source of and sustains the persons of the Trinity, flowing out to and between them.[30] The "flowing" within and among the persons of the Trinity is (or perhaps is mediated by) a fourth thing, *minne*: ". . . God is within all and entirely unenclosed, that is in the eternal enjoyment of himself, and in the dark power of his father, and in the wonder of the love (*minnen*) of himself, and in the strong overflowing flood of his holy spirit" (*Letter* 22:102-7). [. . . God binnen al es ende al onghesloten, dat es in die eweleke ghebrukenesse sijns selues, Ende inde deemstere cracht sijns vader, Ende in die wondere der Minnen sijns selues, Ende inde clare ouervloedeghe vloede sijns heilichs gheests.] In *Letter 20* Hadewijch uses the same motif to discuss the nature of *minne* and how it gathers loving souls to be one with her:

> That nature from which true minne emerges has twelve hours, which stir minne out of herself and bring her back again into herself. And as minne then comes back into herself, she brings [back] those for whom the nameless hours had stirred out: a seeking mind, a desiring heart, and a loving soul. And when minne brings these [in], she throws them into the abyss of the strong nature from which minne is born and fed. Then the nameless hours come into the unknown nature. Then minne has returned to herself and enjoys her own nature beneath, above and all around her. (*Letter* 20:1-14)
>
> [Die nature daer gherechte minne vte comt die heuet .xij. vren die de Minne berueren vte hare seluen Ende bringhense weder in haer seluen. Ende alse Minne dan weder comt in hare seluen, soe bringhet si daerse de onghenoemde vren omme vte hadden gheiaghet, Dat es: een soekende sen, Ende ene begherende herte, Ende ene minnende ziele. Ende

> alse Minne sese bringhet, soe worpt sise inden abis der starker naturen, daer Minne vte gheboren es ende gheuoedet. Dan comen die onghenoemde vren inde onbekinde nature. Dan es Minne te hare seluen comen Ende ghebruket hare naturen beneden hare ende bouen hare ende al omme hare.]

In fact, the internal life of the Trinity seems identical to the dynamic reciprocity of *minne*. Similarly, the "nature," "abyss," or "ground" of God seems to be *minne*-nature, because the ground in which *minne* "enjoys her own nature" is the source of all and the end of all—including *minne* as it or she exists beyond that ground.[31]

Minne is also the essential nature of the human person—or at least some human people. Both the internal economy of God and the structure of the human person are defined by the inward and outward "flowing" of *minne*; or, as Hadewijch often puts it, by the fully reciprocal "demand" of *minne* by *minne*.[32] Within God, the Father demands *minne* (i.e., the unifying ground of God) from both the Son and the Holy Spirit, and they demand it equally of him; likewise the true lover demands the unity of *minne* from God (*Letter* 30; *PM* 1:59-62). The reciprocal demands subdue all parties: "If anyone allowed *minne* to conquer him, / he would then conquer *minne* completely" (*Letter* 19:6-7). [Die ghenoech der Minnen woude wiken, / Hi soude noch Minne verwinnen al.] Within God, *minne* conquers all; likewise God (as *minne*) conquers the true servant of *minne,* and she him.[33] This is a fully reciprocal exchange of "all for all" that perfects both (e.g., *PS* 17:55-60). The perfected person then lives the life of the Trinity; they are not merely "like" the Trinity, they are part of it because they participate in the mutual exchange of *minne.* Indeed, they are the Trinity just as much as the Trinity is itself, because the "flowing" reciprocal demands of *minne* are Trinity's essential nature (e.g., *Letter* 17 and *PM* 4). The perfect servant of *minne* is in God's "being" (*wesen*), and their existence in God is matched by his existence in them (*Letter* 6:29-35, 337-43; *Letter* 18; *Letter* 19:37-45; *Letter* 29:85-91; *Letter* 22:110-13). The reciprocity of *minne* is life of unity: Hadewijch terms this *ghebruken,* "enjoyment," and *ghenoeghen,* "satisfaction."[34] Those who achieve the enjoyment that is *minne* achieve a pure, unfallen state (*Visions* 1:278).[35]

MINNE-NATURE AND THE NATURE OF TOUCH AND TASTE

Hadewijch's concept of *minne* governs the way she uses language of taste and touch. Indeed, she actually labels *minne* "touch": ". . . nothing can dwell in Love, and nothing can touch her except desire; and her most secret name is Touch, and that is a nature that springs from Love herself. For Love is always desiring and touching and devouring in herself. Nevertheless she is in herself entirely perfect" (*Letter* 20:64-70). [. . . Gheen dinc wonen en mach in de Minne noch hare en mach gheen gherienen sonder begherte;

ende hare verborghenste name dat es ghereinen; ende dat es ene natuere, die uter Minnen selve springhet. Want Minne es altoes begherende ende ghereinende ende terrende in haer selven. Noctan es si in haer seluen al volmaect.] "Touch" is the endlessly dynamic "nature" of divine *minne* that is (or is in) the abyss or ground of God. She seems to name it "Touch" because to her that signifies both the act or state of immediate contact and the act of stirring or stirring up. It is significant that in this passage she does not use the words *rueren* or *berueren,* the Middle Dutch words with the primary meaning "to move, stir; to affect (the emotions)." She does sometimes use these words (see, e.g., the passage from *Letter 20* cited above, pp. 80-1). However, *rueren* and *berueren* lack any strong overtone of bodily contact, and I suggest that she prefers words for physical touching, *gherenen* and *gheraken,* because they allow her to characterize the *minne* within the Trinity and between the Trinity and the human person as a "stirring up" that is immediate and reciprocal.[36]

By means of *minne*-nature, which is itself "Touch," the true lover is "stormily touched" (*PS* 21:30) [van storme gherijnt]:"He whom she touches in [*or* with] her nature / must always be / in restlessness / and in great disgrace" (*PS* 24:14-17). [Dien si gherijnt in hare natuere / Dat hi in ongheduere / Altoes moet sijn / Ande in groter onghenade.] Those whom *minne* touches she stirs up into her own restless dynamism, which Hadewijch ascribes to the act of touch itself. The servant of *minne* feels this restless both as endless enjoyment (*ghebrukene*) of *minne* and as equally endless painful desire for her (e.g., *Letter* 20, *PM* 16).[37] Perfect satiety and perfect hunger together are the nature of *minne*:

> . . . those who are born of *minne*
> and are chosen to her nature,
> they therefore spare no pain,
> they live always in holy torment.
>
> He whom high[er] *minne*'s nature touches
> always gladly takes pains,
> as well appears in his deeds:
> he thinks them ever unrefined.
> . . . .
> Satiety and hunger, both in one
> are the fief of lavish *minne,*
> as is ever well known by those
> whom *minne* has touched with her nature. (*PS* 33:13-28)

[Maer die van minnen sijn gheboren, / Ende te haerre natueren sijn vercoren, / Sine sparen ghene pine daer voren, / Si leven altoes in heyleghen thoren. // Dien hogher minnen natuere gherijnt, / Hi es die altoes gherne pijnt, / Als ane sine werken wale scijnt: / Het dunct hem emmer onghefijnt. // . . . // Sat ende hongher, beide in een, / Dat es der vrier minnen leen, / Als ye den ghenen wale scheen / Die minne met haerre natueren ghereen.]

Hadewijch uses language of touch to emphasize her idea that, because *minne* is intrinsically dynamic and turbulent, so must the true lover of *minne* be, and so must the bond between them. One major manifestation of this is the alternating presence and absence of *minne,* which is a pervasive theme in Hadewijch's writings (see, e.g., *PS* 33). Sometimes *Frau Minne* absents herself from her lover, apparently breaking the bond of *minne*—but only apparently. The true lover of *minne* knows that her absence is also a presence, because it is an aspect of her essential nature, which is fluctuating: "to lose one's way in her [*minne*] is to touch her closely; / to die of hunger for her is to feed and to taste" (*PM* 13:2-3). [Jn haer verdolen dats na gheraken; / Om haer verhongheren dats voeden ende smaken; . . .]

Language of touch also helps Hadewijch articulate another idea: the relation between *minne* "here," in people born of *minne,* and *minne* "there," in the divine economy, is fully reciprocal. Hadewijch frequently writes that her narrator touches and is touched by *minne.* God touches the true lover with *minne*'s totality (*gheheeheiden; Letter* 29:66); *minne* touches the true lover with her noble attributes (e.g., fidelity, *Visions* 4:50-58; service, *Visions* 13:132), her depth (*diepheit; Letter* 18:77) or deep touch (*diepen gherinne; PS* 25:35), and with her nature (*natuere; PS* 24:14, 33:17, 33:28). Hadewijch also says the reverse: the narrator desires to touch *minne* (*PS* 43:7; 43:35), and she does touch *minne* (*Letter* 19:12, *Visions* 13:12), *minne*'s nature (*nature; PS* 4:51) and ground (*gront*; *Letter* 18:77), and the being of the Trinity (*wesen; List* 58-60).

I have already argued that reciprocity is central to Hadewijch's concept of *minne,* and that she discusses the internal economy of God as the fully reciprocal demand of *minne* by *minne,* and ascribes the same reciprocal dynamism to the human person. It is a major theme throughout her writings; it is no coincidence that the biblical verse she cites more than any other, by far, is Song 2:16: "you are all mine, love, and I all yours."[38] Of course, the reciprocity of divine love was a well-worn theme by Hadewijch's day. It was especially prominent in Cistercian writings; but among the monks it was a strongly "bridal" concept that was tightly bound to interpretation of the Song. It articulated a tender interchange between God and a soul subordinate and distinct, and, if God fully satisfies the soul with his love, he for his part is only satisfied *with* that loving soul. Hadewijch's idea of reciprocity is fiercer and more radical: *minne* is the turbulent mutual demand of equals in which *both* are fully satisfied (*ghenoeghen*). Her emphasis on full reciprocity and equality follows from her conception that *minne* is a dynamic nature or essence, a "something" that is the "ground" of both God and the human person. Quite unlike Bernard of Clairvux, who taught that love of God was a reciprocal relation of wills, Hadewijch teaches that reciprocity is an *ontological* relation of essences. (The role of the will is significant, but it is not the end of the matter.)

When Hadewijch writes that we taste or touch *minne,* or it touches us, I think we must understand this as *minne* touching *itself.* This is partly because, as we have seen, Hadewijch calls *minne* itself "touch"; but it is also because she seems to teach that *minne* is an nature, essence or "something" within God that is also within the human person—at least a chosen few. Hadewijch appeals to the blood-relation of "nobility" to claim that an elite are predestined and enjoy immediate ontological participation in God here and now. Those "born of *minne*" can enjoy an immediate contact with *minne* because the *minne* that is God or within God can "touch" and stir up the *minne* that they are or that is in them.[39] Those "born of *minne* . . . and chosen to her nature" are "touched by the nature of *minne*" (*PS* 33:13-14). [Maer die van minnen sijn gheboren, / Ende te haerre natueren sijn vercoren, / . . . // Dien hogher minnen natuere gherijnt]. The nobility of *minne* are rewarded with "the enjoyment of powerful taste" (*PS* 20:40) [ghebrukene van weldeghen smaken].

Hadewijch's guiding assumption, clearly, is that "only like knows like." *Minne* is only sought and known by *minne;* only *minne* will satisfy the true lovers of *minne,* because they have *minne*-nature. In one of her letters she writes:

> I greet you, dear, with the love that is God, and with what I am and God is somewhat, . . . one must seek everything by means of itself: strength by means of strength, knowledge by means of knowledge, riches by means of riches, love by means of love, all by means of all, and always like by means of like: this can satisfy him and nothing else. Love, that alone is the thing that can satisfy us, and nothing else. . . . (*Letter* 7:1-12; cf. *PM* 12:1-8, where Hadewijch uses almost these exact words again)
>
> [Ay ic gruete v, lieue, metter minnen die god es, ende met dat ic ben ende dat god yet es. . . . Ay lieue, alle[n] saken sal men met hem seluen soeken: Cracht met crachte, liste met liste, Rike met rike, Minne met Minnen, Al met allen, Ende emmer gheliken met gheliken: dat mach hem ghenoeghen ende anders niet. Minne dat es die sake allene die ons mach ghenoegh doen ende el en ghene; . . .]

Hadewijch is not afraid to draw the corollary: only the *minne* in these chosen few can satisfy divine *minne.* They "feel with God in God" (*Letter* 28:258-59) in a reciprocal relation of touch: "Soul is a way for the passage of God into his depth from his liberty, and God is a way for the passage of the soul into her liberty, that is, into his ground, which cannot be touched, without touching with her depth. And as long as God does not belong to the soul in his totality, he does not truly satisfy it" (*Letter* 18:73-78). [Siel es een wech vanden dore vaerne gods in sine vriheit van sinen diepsten; ende god es een wech vanden dore vaerne der zielen in hare vriheit, dat es in sinen gront die niet gheraect en can werden, sine gherakene met hare diepheit.] Union with God is a fully reciprocal exchange of *minne*. The mutual "passage" in which the depth of the soul "touches" the ground of God is possible because both alike are *minne,* as is the touch itself. *Minne* knows *minne* by means of *minne*. Hadewijch rings on this idea in one of her poems:

> Oh Love, were I Love,
> and to Love you, Love, with Love,
> oh Love, for Love give that Love
> which Love fully recognizes as Love! (*PM* 15:49-52; translation from Murk-Jansen 1992, 124, slightly modified)
>
> [Ay, minne, ware ic minne / Ende met minnen minne v minne! / Ay, minne, om minne gheuet dat minne / Die minne al minne volkinne.]

As with language of touch, Hadewijch seems to use language of taste to emphasize that those born of *minne* enjoy an immediate relation to God—and a fully reciprocal one. According to Hadewijch, we taste many things,[40] but especially God and *minne*—that is, their attributes, including those hidden and secret;[41] divine *minne* itself, what *minne* is, her being or her "nature," or her secret judgments.[42] The true lover tastes "nameless hours"—a cryptic phrase that seems to signify the alternating consolation and suffering that is the paradoxical nature both of tasting her and of *minne* herself :

> Love's way of acting is unheard of,
> as he well knows who has experienced it.
> For she withdraws consolation in the middle;
> he cannot endure,
> whom Love touches;
> he tastes many nameless hours. (*PS* 5:15-20)
>
> [Der minnen pleghen es onghehoert, / Als hi wel kint die hevet becoert. / Want si in midden den troest testoert; / Hine can ghedueren / Dien minne ghereact; / Hi ghesmaect / Vele onghenoemder uren.]

The taste of her nature is perhaps distinct and prior to the touch:

> Before anyone is suckled by Love,
> he suffers great bewilderment; before he touches
> there he tastes
> the nature of Love. (*PS* 5:38-42)
>
> [Eer selc van minnen wert ghesoghet, / Doghet hi grote avontuere, Eer hi gheraect / Daer hi ghesmaect / Der miien natuere.]

The chosen "born of *minne*" know *minne* (divine *minne* in God) immediately by means of the *minne* in them that they are: they know how *minne* tastes.

> Oh noble one full of reason, where may you rest?
> Yes, noblest of all creatures,
> chosen to the nature of *minne*
> in enjoyment of powerful taste:
> powerful *minne* shall reward
> new substance, bliss, blossoming at all times.[43] (*PS* 20:37-42)
>
> [Ay, edele vol redenen, waer mogdi dueren? / Ja, edelste alre creaturen, / Vercoren ter minnen natueren / In ghebrukene van weldeghen smaken: / Nuwe materie, bliscap, bloyen alle uren / Sal gheweldeghe minne orsaten.]

Minne gives herself to them fully in a taste that renews (*PS* 1:101), and by tasting *minne* they attain full satisfaction (*ghenoeghte*), which is the reciprocal enjoyment of *minne* by *minne*:

> The taste that fidelity gives in Love
> . . . .
> is heavenly satisfaction, free,
> full, without deficiency:
> "you are all mine, love, and I all yours." (*PS* 34:41, 45-7; cf. *PS* 33:33-36)
>
> [Die smake, die trouwe in minnen ghevet, / . . . / . . . hets helmsche ghenoechte vri / Te vollen, sonder ghebreken: / "Du mi al lief ende ic al di."]

Hadewijch is clear that the taste and touch of *minne* give knowledge, which she explicitly distinguishes from the knowledge accessible to intellect or reason (*redene*). She especially favors language of taste to talk about such knowledge. The fruits of union with *minne* and God are a taste (*smake*) that gives knowledge (*bekinnesse, bekinnen*).[44] In her book of *Visions,* the narrator is told by God "you have tasted and received me outwardly and inwardly; and you have understood the ways of union that wholly begin in me" (*Visions* 8:98). [Nu hevestu mijns ghesmaect ende ontfaen van buten ende van binnen; ende du hevesverstaen die eneghe weghe die gheheel in mi beghinnen.] The narrator desires "fully to enjoy [her divine] lover, and to understand and taste him in all full attainment" (*Visions* 7:22). [Ic begherde mijns liefs te vollen te ghebrukene ende te bekinnenne ende te ghesmakene in allen vollen ghereke: . . .] She is told by a divine messenger that the "face" of God "fully teaches divinity and fruitfulness and gives richness of all enjoyment of all full taste of cognition . . ." (*Visions* 8:21) [anschine . . . vollei det ende leret diuiniteit ende vroetheit ende rijcheit gheeft aller ghebrukenessen van allen vollen consteleken smake]. As the true lover thus tastes and knows, likewise *minne* and her heavenly servants taste and know her (*Visions* 4:120-29).

This knowledge is gained through *minne,* and not otherwise. Hadewijch makes this point clear in *Letter 18*. Just after a passage where she emphasizes that the soul has an essence that enjoys full reciprocity with God and touches him immediately (which I have quoted above, pp. 84-5), Hadewijch gives one of her more extended treatments of the relation between reason and *minne,* using the motif of love's two "eyes" (*Letter* 18:174-96).[45] She simultaneously seems to distinguish between the "sight" of reason and the more immediate taste and touch of *minne*. The soul has a "power of sight," which is charity (*caritate*). Charity has two eyes, love and reason (*minne* and *redene*). Hadewijch explains how both are needed, but then at the end shifts from language of sight to language of touch and taste: one should gaze at and inspect God intensely, such that "the eyes of your desire both together remain fixed on the countenance of your beloved with piercing nails of burning touches that never cease" (*Letter* 18:179-82) [dine gheenighde oghen dijnre begherten bliue ane hangende in dat anschijn dijns liefs Metten dore gaenden naghelen der berrender gherijnnessen die niet en cesseren]. Then *minne* "makes him taste her so sweetly that he forgets everything on earth. . . ." (*Letter* 18:195-97) [Minne altoes haer seluen hem soe suete smaken doet, Dat hi al dies verghet dat in ertrike es. . .].

This letter is a bit confusing, because *minne* seems to act on both the soul's "eyes," *minne* and reason. But Hadewijch makes her overall point clear: when we taste, touch and are touched by *minne,* we know what God is: "Reason cannot see God except in what he is not" (*Letter* 18:82-83). [De redene en can gode niet ghesien sonder in dat hi niet en es; . . .] That is not to say reason is unimportant: "When reason abandons itself to love's wish, and love consents to be forced and held within the bounds of reason, they can accomplish a very great work" that exceeds even wisdom: the enjoyment of *minne*; and this is known by those who feel it (*Letter* 18:95-97). [Alse redene dan valt in begherten van Minne ende hare Minne dwinghen laet ende benden ten steke der redenen, soe vermoghense een ouer groet werc: dat en mach nieman leren sonder met gheuoelne. Want wijsheit en minghet hare daer toe niet, . . . die alle wesen verboghen es, sonder ghebrukene van Minnen.] Hadewijch strongly emphasizes that without reason no one can fully serve *minne* or enjoy union. Reason both leads the true lover to the depths of *minne* and continually reminds her that she is wanting; reason thus maintains the dynamically oscillating torment that Hadewijch insists is the true nature of *minne*. Reason is "the physician of *minne*" *PS* 25:83) [der minnen fisisijn] who heals by wounding. However, reason cannot attain to all that *minne* can. True *minne* is touched and tasted by avoiding paths and ways: *minne* will bear the true lover "beyond all the ways people can think of / . . . / [and] reason cannot understand / how *minne* by *minne* sees to the depth of the lover" (*Letter* 19:11-20) [buten allen weghe van menschen sinnen, / . . . / want redene en mach begripen niet, / hoe Minne met Minnen lief doer siet, / Ende how minne in allen leuet vri; . . .]. Indeed, with the mind (*sinne*) one cannot speak of *minne*; one must speak with the soul (*Letter* 19:27-30). The precise arrangement of the soul and *minne* is not clear, nor is that between *sinne* and *redene*. Although apparently somehow distinguishable, *sinne* and *redene* seem closely related as forms of discursive understanding. Nevertheless, Hadewijch's point is clear: *sinne* and *redene* are distinct from *minne,* which alone can feel, know, and be one with *minne*. Reason is necessary but insufficient for union with *minne*. It is the part of the human person (or the act of discursive understanding) that can help know *minne* but cannot itself be one with it. One reason Hadewijch insists that *minne* and reason are complementary is to criticize the sentimental and self-serving who seek to serve and attain *minne* only in order to gain sweet feelings of consolation; they seek their own will, not the divine will of *minne*. Hadewijch insists that without reason we cannot know the true paradoxical harshness of *minne*—her restless, agitating, and painful touch.

What a person knows by reason is not superseded but is exceeded by the *minne* they know immediately by means of *minne* (*Letter* 11:44-46). Hadewijch writes that beyond the seven gifts and tokens of *minne,* there is an eighth reserved to the chosen few born of her: "And the eighth [token of *minne*] is the touch of enjoyment, which does away with everything that pertains to reason, and lover falls [as] one in[to] lover" (*Visions* 13:179-82). [Die vij gauen sijn vij tekene der minnen. Ende dat achtende es ghrijnnesse van ghebrukene die al af doet datter redenen behoert, ende lief in lief een valt.] At this point "*minne* practices *minne* with *minne*" (*PS* 20:76) [*minne der minnen met minnen pleghet*] in an "embrace" that teaches but escapes (rational) understanding. Likewise, "she [*minne*] causes us to long for her with desires / and taste without knowing her being" (*PS* 22:31-32) [si doet met begheerten na hare haken / Ende sonder kinnen hare wesen smaken]. The true lover of *minne* knows the being or essence of *minne*—knows *minne*-nature—by means of *minne,* with a direct contact, not by discursive knowing: ". . . we are made for this, / to know with *minne* how *minne* tastes, . . ." (*PM* 12:22). [Nadien dat wiere toe sijn ghemeact / Te wetenne hoe minne met minnen smaect. . . .] The true lover, "chosen to the nature of *minne*" (*PS* 20:40) [*vercoren ter minnen natueren*] tastes by means of *minne* because she is *minne*: "I could taste humanity and God in one knowledge, which no human being does unless he were already as God and wholly was who our *minne* is" (*Visions* 14:161-65). [. . . Ic toe ghecoren was dat ic mensche ende god in eenre const smaken soude, dat nie mensche doen ne mochte hine ware al alse god ende altemale was die onse minne es.] The true lover tastes what no person can never know intellectually, for "*minne* permits the soul to taste her secret judgements, which are deeper and darker than the abysses [of God]" (*Letter* 20:37) [de Minne der zielen ghevet te ghesmakene hare verhoelne ordele die diepere ende donckere sijn dan die afgronde]. Thus she can understand "that hidden word of himself [God] out of this depth" (*Visions* 12:145-48; "depth" ascribed to the narrator at 12:167-69) [{sie} dat verhoelne waert hadde verstaen van hem seluen vte diere diepheit]. The hidden word "no alien can understand, / except he who has already felt it by taste / and in great noise received silence" (*PS* 4:14-18; see also *Letter* 18:187) [niemand vremders en mach verstaen, / Dan diet van smake ghevoelt al hevet / Ende in hoech gheruchte scilentie ontfaen].

CHRIST, *MINNE*, AND BODILY LANGUAGE

We have just encountered a significant phrase that Hadewijch echoes a number of times: the chosen few "taste humanity and God in one knowledge" (*Visions* 14:161-62) [mensche ende god in eenre const smaken]; by means of *minne* they taste "human and divine *minne* in one single being," which is "the taste of the undivided nature" (*Visions* 14.80-84) [menscheleker ende godleker minnen in enen wesene . . . natueren smake onghesceden]. Obviously, this is christological language. She uses it not only to emphasize the orthodox teaching that the human Christ was fully God, but to emphasize that we have access to both aspects of his undivided nature. We have immediate access to both his natures in the Eucharist and, more profoundly, we can be part of their integral cohesion in the bond realized by *minne*. The bond of *minne* is not reserved for a spiritual state that is deferred to the afterlife, but can be realized here and now. Hadewijch uses language of taste and touch to articulate the immediacy of this union. She does not do so in a way that implies the operation of a second set of spiritual senses that allow us to know we know *minne*, God, and Christ in a "spiritual" mode or manner or by means of a special, second set of "spiritual senses"; rather, she implies that we know divine things by means of a single sensorium.

Hadewijch repeatedly insists that the true lover of *minne* must love and suffer in the real physical world, as Jesus did, in perfect service of *minne* (*Letter* 15:29-35, *Letter* 6:350-68; *PS* 30:25-30). To be and to know in the body is to be and to know as Jesus did, which is perfectly. Hadewijch assumes that Jesus Christ knew by means of his bodily humanity. That idea is not easy to square with the assumption that God knows all. The problem was recognized already by early Christian thinkers, and we have already seen that Bernard gets tangled in this problem when he insists that Christ experienced, and therefore knew from experience just as all human beings do. But the problem does not seem to trouble Hadewijch. She assumes that by serving *minne* perfectly, as Christ did, the true lover learns and knows *minne* just as Christ did, in the world and by means of their physical body.

Hadewijch goes further, and projects this assumption into the ontological realm of God's inner life: those who serve *minne* perfectly achieve a status equal to Christ's. This is a corollary to her idea that the life of serving *minne* here in the world is continuous with the life of *minne* within God the Trinity. The whole point of the *Book of Visions* (and a significant theme throughout Hadewijch's writings) is to show how the narrator is perfected and achieves the "full-grownness" that grants status equal to Christ's. Those who become full-grown are one with *minne* (and therefore God) in *minne* and to "live God and human" (*Letter* 18:7-8) [gode ende den mensche leuen]. To do so the true lover must both serve and suffer in the world—perform "works—and also seek nothing but *minne* or remain "undivided" or "withheld" and let everything be, both high and low (*Letter* 17:67-74; *Letter* 18:53-57).

A clear illustration of this is *Visions 12*, where the narrator becomes one with God/*minne* while wearing a "robe of undivided will" adorned with her virtues. Among these defining qualities and acts is "peacefulness, [which] showed and bore witness to her, as pleasing in appearance and beautiful, and as possessing knowledge of the total embrace and thoroughly piercing kiss, and of all honor and all encounter as lover shall practice [with] lover in love" (*Visions* 12:114-15) [vredeleecheit . . . toende hare ende orconde goet ghelaet ende scone ende constech in ghehelene behelsene ende in eneden doregaenden cussene ende in alre eren ende in allen hanteerne also lief liefs in minnen pleghen sal. . . .]. She goes on to say that to achieve this union (which, note, she characterizes with tactile language) is to live Christ's life in the Trinity and in the world. The narrator's "peacefulness"[46]—a token of nobility and freedom—shows that she was born with Christ eternally from God; born from a woman with him; grew up with him in pain and poverty and compassion, "nourished from within and without from the other." She suffered and died with him and

> rose with him and stood one with him before his Father, and there with him acknowledged his Father to be his Father, and with him [acknowledged] him to be Son, and with him acknowledged the Holy Spirit to be Holy Spirit, and with him she knew them [to be] entirely like him, [knew them to be] one and [knew] the being in which they are one. To this all her peacefulness bore witness for her, that she has acted thus and that later she shall act, perfectly his, truly with minne in minne. (*Visions* 12:126-34)

> [. . . ende met hem opuerstont ende met hem te sinen vader i voer ende daer met hem sinen vader vader bekinde ende hem met hem sone ende met hem den heilighengheest heilech gheest bekinde ende met hem ghelijc hem kinne sise al i ende dwesen daerse met i sijn. Dit orcont hare al vredeleecheit, dat sijs dus gheploghen heeft ande voert volmaecteleke sijns wel met minnen in minnen pleghen sal.]

That word "later" (*voert*) seems to mean that living *minne* within divine *minne* is temporally distinct from living Christ's suffering life here. In *Letter 17* Hadewijch distinguishes living *minne* within *minne* from suffering service in the world as an alternating sequence of states: within enjoyment (*ghebruken*), one must perform no acts, virtuous or otherwise, no particular works that might assist neighbor, but "remain constantly in the enjoyment of *minne*. But when this enjoyment falters or sinks away, then one must perform all three of the forbidden works by justice and by right" (*Letter* 17:87-90). [. . . {S}lach ouerslach in ghebrukenessen van minnen. Mer in dat faelieren Ende in dat sincken van ghebrukene, dat werctmen wel alle drie de verbodene werke bi scoude ende bi rechte.] To serve *minne* is to be human and needy, "but in the enjoyment of *minne* one becomes God, mighty and just" (*Letter* 17:96-97). [Mer in ghebrukene van Minnen es men god worden moghende ende gherecht.] Similarly, in *Letter* 6 she writes that to bear the cross and suffer with the son of God is to wait for *minne* to show herself:

> That cross that we must bear with the son of the living God is the sweet suffering exile that one bears for true love, ... By this she [Love] shows herself so unreservedly to the loving one that she makes him go out of himself; and robs him of heart and mind, and causes him to die to himself and live in the custom of true Love.
>
> But before *minne* thus overflows [herself], and before she takes a person so much out of himself and touches so closely with herself that he is one spirit and one being with her [and] in her, a person shall offer her both beautiful service and wretched, exiled living: beautiful service in all virtuous works, and wretched, exiled living in all obedience." (*Letter* 6:350-68)

> [Dat cruce dat wi metten leuende gods sone draghen selen, dat es die soete ellende die men om gherechte minne draghet, . . . Daer met toent si haer seluen den minnenden soe temel, dat sine vte hem seluen doet gaen, Ende roeft hem herte ende sen, Ende doeten steruen Ende leuen in pleghene der gherechter minnen.
>
> Mer eer Minne dus ouerbrake waert ende eer si den mensche soe sere vte hem seluen nemt, ende soe na met hare selven gherijnt dat hi een gheeste ende een wesen si met hare in hare, soe sal de minsche beiden scoenen dienst ande ellendich leuen: scoenen dienst in allen doechdeleken werken, ende ellendich leuen in alre ghehorsamheit: . . .]

In *Visions 13* Hadewijch suggests that this sequence corresponds to life in and life out of the body: Mary tells the narrator that "if you wish to have ampler enjoyment, as I have, you must leave your sweet body here" (*Visions* 13:241-42). [Sich wiltuus alsoe voert meer ghebruken alse ic Soe moestu dinen sueten lichame hier hebben.]

In these passages, the unifying touch of *minne* seems distinct from serving her as the man Christ did, in the suffering and sensing body. Because the two alternate, they seem mutually exclusive: either one lives divine *ghebruken* beyond all things, or one lives in suffering human service to *minne*. But Hadewijch repeatedly denies this implication: both must coexist (e.g., *Letter* 6, *Letter* 30). We perhaps better understand union in *ghebruken* and *ghenoeghen* not as second and higher in a sequence, but as a persisting element or aspect in those born of *minne*. In them the element of *ghebruken* and *ghenoeghen* endlessly fluctuates by its very nature, and by its nature inflicts enjoyment and suffering alike, which are both are equally presence and absence. The apparent distinction between enjoyment and suffering dissolves: bodily service to *minne* in this life is an aspect of perfect union with *minne*, and the reverse. Hadewijch calls this paradox "living God and man" and "living human and divine *minne* in one single being." She considers it inherent in the nature of *minne*, and also in the nature of Christ and the Trinity:

> . . . one must Love and know
> the humanity just like the divinity.
> With the live coal, Love sets the two aflame;
> with the fire, she burns them into one,
> . . . .
> With the dew the burning is soothed,
> and salved with a single breeze.
> The delight and the madness of Love
> then throws him into the deepest flood
> which is groundless and always lives
> and with the living one[ness] of the three gives
> God and man in one single Love:
> this is the Trinity beyond all mind [thought]. (*PM* 16:183-96)
>
> [. . . men die menscheit moet minnen / Ghelijk der godheit ende bekinnen. / Metten cole ontstect si hen tween; / Metten viere verbernt sise een, / . . . / . . . / Metten dauwe wert die brant ghesacht, / Ende ghesaluet metter enegher locht. / Die welheyt ende die orewoet / Worpse dan in die diepste vloet, / Die grondeloes es ende altoes leuet / Ende metten leuenne hen drien een gheuet / Gode ende menschen in ene minne: / Dits drieheit bouen all sinne.]

Hadewijch does not clearly distinguish the life of *minne* "here" from the life of minne "there." In *Letter* 6, she writes: "With the humanity of God you shall live here in labor and in exile, and with the powerful eternal God you shall love and rejoice inwardly with a sweet confidence, and the truth of both of these in a single enjoyment (*ghebruken*)" (*Letter* 6:117-21). [Metter mescheit gods suildi hier leuen in aerbide ende in ellenden Ende metten moghenden eweleken god suldi Minnen ende Jubileren van binnen met enen sueten toeuerlate Ende haere beider waerheit es een enich ghebruken.] The true lover who lives Christ's life of *minne* achieves perfect union with *minne* here, in time, and also participates ontologically in the eternal Trinity, in the *diepheit* that is the nature of *minne*. The bond of *minne* does not lift a person out of this worldly realm to a separate divine realm of spirit, it incorporates them into the ground of both.

EUCHARIST AND SOMATIC LANGUAGE OF INDISTINCTION

Like living Christ's life of *minne*, partaking of the Eucharist helps bring the true lover to full-grownness in *minne* and maintains the eternal bond and action of *minne* that is everywhere and always one. Indeed, Hadewijch uses eucharistic language to emphasize her most radical teaching on union: that the bond of *minne* is union "without difference" (*sonder distinctie*). Hadewijch's concept of union without distinction follows from her teaching on reciprocity. Not only does *minne* allow those born of *minne* to affect God as much as he affects them, but true lovers "become *minne* with *minne*" (*PS* 34:79) [minnen met minnen worden] and can "be God with God" (*Letter* 23:121) [god met god sijn]. We have already seen that Hadewijch uses language of taste and touch to articulate this reciprocity, and it is no coincidence that she appeals to these same senses, with the added notion of assimilation, to articulate her concept of union without difference.

Before exploring Hadewijch's eucharistic language a little further, I would like to clarify how I am using the concept of union without distinction and discuss its close relation to theological anthropology. Bernard McGinn has argued that there are two basic concepts of union in medieval theological texts, which we can summarize with the Latin terms *unio indistinctionis* and *unio spiritus*. Even though they overlap to some degree and often both occur in a single text, they remain distinct types. *Unitas spiritus* carefully preserves a distinction between the human person and God; it corresponds, roughly, to the anthropology of image and likeness, which was dominant up until the twelfth century. *Unio indistinctionis* is the idea that a person and God can be one without distinction, because they are not merely "like" but are in some essential respect the same thing. This follows from the idea that human beings, like all creatures, emanated from God, and so share some essence with him that transcends the distinction between them. The shared essence sustains humans and all created things, and it allows humans (and, perhaps, all creation) to "return" to and "be" God. *Unio indistinctionis* therefore often correlates with a more exemplarist anthropology that emphasizes that we participate in God, or even "virtually exist" in God.[47] These ideas have roots in patristic teachings, but they become prominent in Western Europe only after the twelfth century.

Using McGinn's general distinction, we can observe that Bernard of Clairvaux emphasizes that union with God is *unio spiritus*: a moral and psychological transformation that, however immediate, maintains a firm distinction between lover and beloved, subject and object. We have seen that Bernard uses langauge of taste and touch to discuss this intimate but discrete relation: for example, as an *adhaesio* that is never identity, as a mutual eating in which neither agent nor patient is consumed or assimilated, and as a tasting that is immediate but maintains distinctions. Union with God (and the desire that leads to such union) is a form of "likeness." We can observe, in contrast, that although Hadewijch sometimes discusses a similar conception of *unio spiritus* based on likeness, she also discusses *unio indistinctionis,* which is based of her notion of the shared "nature" or "being"of *minne.*

Hadewijch teaches that those born of *minne* can be satisfied by God's "pure nature" as he is within that nature, which is *minne.* I have shown how she uses language of touch and taste to emphasize that this "satisfaction" is fully reciprocal (see, e.g., *PM* 12:61-65). I have already cited a passage where she uses language of touch to write about both union of spirit and union of being: "*minne* overflows herself and . . . takes a person so much out of himself and touches so closely with herself that he is one spirit and one being with her [and] in her . . ." (*Letter* 6:361-64). [. . . Minne dus ouerbrake waert ende . . . si den mensche soe sere vte hem seluen nemt, ende soe na met hare selven gherijnt dat hi een gheeste ende een wesen si met hare in hare. . . .] Note the idea of reciprocity in their actions: as the person is taken out of himself, so too *minne* "overflows herself." In the *Book of Visions,* Hadewijch writes that the reciprocal "flow" of *minne* overcomes distinctions: she refers to those who "experienced human and divine *minne* in one single being," who "understand . . . the taste of the undivided nature, with total divinity thoroughly flowing, and, all one, thoroughly flow back again into divinity" (*Visions* 14:82) [becoerde menscheleker ende godleker minnen in enen wesene, ende . . . verstaen . . . natueren smake onghesceden metter gheheelre godheit dore vloyende, ende alle een weder dore vloye in die godheit]. When the lover is one being with *minne* and in her, it gives full satisfaction to both (*Visions* 1:399-400).

In several passages that are heavy with somatic language Hadewijch amplifies this point by combining language of tasting and touching, kissing and embracing—eucharistic references to eating, drinking, and swallowing (in addition to the passages I discuss in this section, see *Visions* 14:163, 14:163, 7:74, 8:98, and *Letter* 22:296). Behind her references to the Eucharist seems to be the brute fact that eating and drinking are actions that require immediate bodily contact and bodily engulfing, and at least imply assimilation. These eucharistic passages emphasize how Hadewijch's language of tasting and touching is profoundly Christological, that those born of *minne* know and are one with God both because they participate ontologically in the nature of *minne* and because they act and sense God in this world in the body, especially in receiving the Eucharist.

For example, she writes that "by all these tokens which I found between him and me in close intercourse of love—as when friends customarily between themselves hide little and show much—one possesses that most in closely feeling each other, and in thoroughly tasting and in thoroughly eating and thoroughly drinking and swallowing each other" (*Letter* 11:20-27).[48] [Ende bi al dien tekenen die ic vant tusschen hem ende mi in na pleghene van Minnen, alsoe alse vriende pleghen deen den anderen luttel te helene ende veel te toenne, datmen alre meest hevet in na ghevoelne elc anders, ende in doer smakene, ende in doer etene, ende in doer drinckene, ende in verswelghene elc anderen.] The "tokens" or gifts of exchange shared with God both mark and seal a bond like that of friendship, and Hadewijch refers to taste, touch, eating, and drinking to emphasize that this bond is intimate and immediate. The reference to the ritual of the Eucharist is also obvious. Hadewijch assumes that in partaking of the Eucharist a person can know and be one with God and perhaps assumes that the ritual legitimizes language of bodily contact and bodily engulfing. But that can only be part of the reason she uses it here, because the phrase "each other" makes clear that she means this sensory language to emphasize that this encounter is immediate and *fully* reciprocal. Hadewijch uses language of eating and drinking, as she uses language of tasting and touching, to articulate her idea that human beings and God share a *minne*-nature that allows them to achieve union. Hadewijch conceives of union as an immediate relation that can ultimately become identity, such that and a person and God are one "without distinction."

She is fully aware that this implies union without distinction. In fact she insists on the point by writing explicitly that to eat the Eucharist is to be eaten and so to be one with *minne*/God. The idea of mutual eating, if not common, was old already in her day; it goes back at least to Augustine, and we have seen that Bernard wrote something similar (*SC* 71:5; see above, pp. 64-5). But neither of these men made use of the eucharistic reference; more significant, Bernard went out of his way to emphasize that his language of mutual eating *did not* mean union of indistinction.

Hadewijch uses the phrase "one without difference" in a well-known eucharistic passage from *Visions* 7, which is dominated by the spatial concept of two bodies touching and engulfing each other:

> Then he gave himself to me in the species of the Sacrament, in its outward shape, as is the custom; and then he gave me to drink from his chalice, in shape and taste, as is the custom. After that he himself came to me and took me entirely in his arms and pressed me to him; and I felt all his members in all her [*minne*'s?] enjoyment, according to the desire of my heart [and] according to my humanity. So I was outwardly satisfied in all full satiety. And also for a short while, I had the strength

> to bear this; but soon, after a short time, I lost that beautiful man outwardly in the sight of his form, and I saw him completely come to nothing and so fade and entirely dissolve into one, that I could not recognize or perceive him outside me, and I could not distinguish him within me. At that time it was to me as if we were one without difference. This was entirely outward in seeing, tasting and feeling; just as one can taste in receiving the sacrament outwardly, in seeing and in feeling outwardly, so can one receive the lover in all full enjoyment of seeing and hearing and carrying away of one in the other. (*Visions* 7:70-99)
>
> [Doe gaf hi mi hem seluen in specien des sacraments in figuren also men pleghet; Ende daer na gaf hi mi drincken vten kelke ghedane ende smake also men pleghet. Daer na quam hi selve te mi, ende nam mi alte male in sine arme, ende dwanc mi ane heme; ende alle die lede die ic hadde ghevoelden der siere in all hare ghenoeghen na miere herten begherten na miere menscheit. Doe werdic ghenoeghet van buten in allen vollen sade. Ende oec haddic doe ene corte wile cracht dat te draghene. maer saen in corter vren verloesic dien sconen man van buten in siene in vormen, ende ic sachene al te niete werdene Ende also sere verdoienede werden ande al smelten in een, Soe dat icken buten mi niet en conste bekinnen noch vernemen, Ende binnen ni niet besceden. Mi was op die vre ochte wi een waren sonder differencie. Dit was al buten in siene, in smakene, in ghevoelne, alsoe men smaken mach van ontfane inden sacramente van buten, in siene ende in ghevoelne van buten, also lief man ontfaen mach in aller vollerj ghenoechten van siene ende van hoerne, van vervaerne deen inden anderen.]

This is a complex passage, but it well illustrates how Hadewijch does not observe any clear distinction between spiritual and bodily oneness with God. Perhaps she intends to preserve distinctions; she does after all qualify the phrase *sonder differencie* with "as if" (*ochte*) and then speaks of its "outwardness."[49] On the other hand, she refers to tasting and touching the Eucharist and Christ the man without using any qualifiers. As these references to "outward" signify the immediacy and full reciprocity of union, so the references to eating the Eucharist and to "dissolving" signify full mutual interpenetration.

Hadewijch's eucharistic formulation of *unio indistictionis* echoes other passages where, as we have seen, she writes about the reciprocal flow between *minne* and the person and the oneness of their being. In *PM 16*, Hadewijch elaborates this eucharistic language for union to include the ideas she condenses in the concept of *minne.* When a person eats the bread they eat Christ, who is *minne,* and so become one with him in the paradoxical ways of *minne:* in enjoyment, in pain, in satisfaction, in separation, in knowledge:

[Love's] bonds bind all seams
In one enjoyment, in one satisfaction.
This is the bond that totally binds,
so that each thoroughly knows the other
in pain, in rest, in madness,
and eats his flesh and drinks his blood,
the heart of the one thoroughly devours the other's heart with storms,
as he who is himself love showed us;
he who gave himself to us to eat
gives that beyond human mind [*sinne*].
Thereby he gave us to know
that *that* was the nearest to *minne*:
to thoroughly eat, thoroughly taste, [and] thoroughly see from within.
(*PM* 16:27-37)

[Dit es die band, die al dat bint / Dat deen den anderen dorekint / Jn pine, in raste, in orewoet, / Ende etet sijn vleesch ende drincet sijn bloet, / Ende dene herte dandere met storme doreueert, / Also hi ons toende die selue es minne; / Dat gheet bouen menschen sinne, / Die ons hem seluen gaf tetenne. / Daer bi dedi ons te wetenne / Dat dat ware die naeste der minnen / Dore eten, dore smaken, dore sien van binnen.]

Later in this poem Hadewijch picks up on the eucharistic reference again to discuss union of essence, this now accenting the trinitarian aspect of union and the identity of *minne* in herself, Christ, and in the true (human) lover (note the shifts in the gender of the pronouns, a technique Hadewijch often uses [Murk-Jansen 1996]):

When he [Christ] embraces [her] in all thought,
she thoroughly suckles and thoroughly tastes.
As Love thus touches the lover,
she eats her flesh and drinks her blood.
Love thus does away with these,
sweetly seduces them both
in a kiss without division;
that kiss unites beautifully
three persons in one sole being. (*PM* 16:116-24)

[Alse hi lieue beueet in allen sinnen, / Si doresughetse ende doresmaket./ Alse minne die lieue dus gheraket, / Si et hare vleesch, si drinct hare bloet. / Die minne diese dus verdoet / Verleidet suetlike hen beiden / Jn enen cussenne sonder sceiden; / Dat cussen enicht scone / Jn enen wesenne .iij. persone.]

The "kiss without division" is the stormy relation of *minne* with *minne*. This is evident if we compare a sequence of lines just before these with an almost identical sequence in another poem:

> [Dew] . . . brings that kiss of noble natures
> and gives rest in restlessness.
> Need so swallows her gifts
> that she must always practice this mode of action. (*PM* 16:107-10)
>
> [Dau . . . / . . . brinct dat cussen der edelre natueren / Ende doetse in ongheduerne dueren. / Die niedt verswelghet so hare gheuen, / Datse dies seden altoes moet pleghen.]
>
> Nature [so] demands [his] nature with storms
> and gives rest in restlessness
> need so swallows all her gifts
> that she must always practice this mode of action. (*PM* 1:53-56)
>
> [Natuere eiscet soe met storme natueren / Ende doete in ongheduerne duernen, / Die niedt verswelghet so al hare gheuen, / Dat si dies seden altoes moet pleghen.]

Note the difference: in place of the kiss, we have the phrase "nature so demands nature with storms." That is, Hadewijch uses language of touch to indicate the immediate relation of *minne* to itself, which is the ground of union with God. *Minne*-nature's restless reciprocal "demand" is the essence of *minne*: wherever it exists, *minne* has a turbulent, endlessly restless relation with itself. In *PM 16*, Hadewijch once again articulates this idea with tactile language: here a touch of the lips refers not only to the loving contact within the Trinity, but also to Christological language of union in Eucharist: "suckling," "eating" and "drinking" lead to a "kiss without division."

My claims about Hadewijch' somatic language of indistinction partly elaborate those made by Caroline Walker Bynum (Bynum 1988, 153-61). She focuses on language of eating and hunger, which are distinct from sensory language, but her point about Hadewijch's basic assumptions applies to other somatic language as well, including language of taste and touch. Bynum emphasizes that Hadewijch uses somatic metaphors because she (Hadewijch) believes "we experience God through all our senses" and with our full humanity (Bynum 1988, 160). It is because God is a physical man as Jesus Christ that we can know and be one with him—and this human physicality is not left behind or negated by the exalted spiritual nature of God. Bynum therefore suggests, without making explicit, my point about Hadewijch's anthropology: Hadewijch observes no clear distinction between the material, physical, bodily person and "spiritual" entities such as *minne* and God because "to Hadewijch, the God who is infinite abyss beyond all language and metaphor is also fleshly humanity—a humanity that suffers and feeds" (Bynum 1988, 160).

Despite my caveats about the "experiential" aspect of Bynum's analysis, I agree with her that Hadewijch finds a warrant for somatic language in the fact that Christ was a physical man, and in the fact that believers have access to him and to God in the Eucharist they taste and eat. Somatic language is based on the idea of "physicality joining physicality," does seem to find grounding in and simultaneously reinforce a piety that emphasizes physical *imitatio Christi* and the Eucharist. Christology and Eucharist are only one aspect of Hadewijch's teaching. So beyond Bynum's point, I have argued Hadewijch has at least one other reason to use somatic language of taste and touch: it helps her articulate her abstract, ontological and trinitarian theology of *minne*. This theology in turn assumes a highly sophisticated anthropology.

My argument implies that Bynum's position is too narrow, insofar as she finds Hadewijch's warrants for highly somatic language in the *practices* of physical piety.[50] I do not deny that such practices profoundly shape thought, including theological reflection; and Bynum would in no way deny that thought shapes practice as well—indeed, her argument depends on this assumption. But in her analysis of Hadewijch (at least), Bynum tends to mute this kind of influence, because she almost completely ignores the conceptually sophisticated trinitarian theology of *minne* that is the heart of Hadewijch's thought and teaching.

I think it is crucial to notice this because otherwise Hadewijch's highly somatic sensory language—language of taste and touch—seems *merely* somatic. Hadewijch's theology of *minne* works in conjunction with her teaching that we know God in the body because he was the physical man Jesus to maintain a "spiritual" pole for her sensory language, while simultaneously locating that spiritual pole in the physical person. Unlike Origen, and even more than Bernard, Hadewijch discusses the ontological aspects of *minne* ("spiritual" aspects—though she does not make the distinction) by appealing to the embodied person. Moreover, she locates the most profound ontological essence of *minne* in the experienced body, such that the union of minne with minne allows those born of minne to be perfect in *this* life and achieve a status equal to or greater than Christ's.[51]

Hadewijch assumes a concept of spiritual sensation as integral as Bernard's, although transposed to a more abstract, ontological register. She uses language of taste and especially touch to talk about our relation to God in a way that implies a human person knows God not by means of a second set of spiritual senses, but by means of a single sensorium. Hadewijch seems to feel no need for a second set of spiritual senses (and so no need for a carefully circumscribed use of sensory language) because she recognizes no *essential* gap between God and the human person (at least the elect born of *minne*). Rather, she asserts that they are one in their shared essence or nature, whose actions and effects she often discusses in highly somatic, sensory language of taste and touch. She has compelling theological reasons to do so, especially her explicit emphasis that we must live as the human Christ, so we will with *minne* and live within the trinitarian dynamism of God. In anthropological terms, we can say that she does not seem to think that our souls are fundamentally distinct from our material body. She assumes that a person is not so much a soul *in* a body, but a soul *and* a body. Therefore, she considers the physical body less an impediment to union with God than the means for achieving it.

Chapter 5

Echoes and Ambiguities

We have already seen that Hadewijch was not the only thirteenth-century theologian to associate taste and touch with union with God: so did Albert the Great. However, he did not share her ideas about the structure of the human person, and his concept of the spiritual senses clearly falls into the dualist and intellectualist family, as is clear from his statements that the spiritual senses are cognitive acts of the intellect alone. Several of the thirteenth and fourteenth century theologians who are generally credited with significant mystical texts shared Hadewijch's ideas about immediate union with God as a kind of latent "virtual existence" in the ground of the divine—Marguerite Porete and Eckhart, to name the two best-known—and you might expect them to have used sensory language in a way similar to her, as well. However, Porete and Eckhart prefer to articulate that idea (or suite of ideas) in language that relies much more on logical and syntactic aporia. Porete uses very little sensory language, Eckhart rather more, but almost all of this is language of sight and light. Indeed, they are singularly uninterested in how our senses and, by extension, our bodiliness allow us to know God—except negatively. Hadewijch's contemporary, Mechthilde of Magdeburg, shares with her some of anthropological ideas and uses sensory language fairly extensively. Yet she, too, heavily favors language of sight and light.[1] Although her usage suggests a concept of the spiritual senses, it does not seem to belong to the same family as Hadewijch's and Bernard's ideas.

Other authors of mystical texts, however, did use significant language of taste and touch and suggest a similarly integral concept of the spiritual senses, at least in some respects. The most striking example is the fourteenth-century Dutch theologian Jan Ruusbroec (1293-1381). He talks about union with God as an immediate touch—for his own reasons, but there is not much question that he was partly influenced by Hadewijch's ideas and writings. Another example is the German Franciscan Rudolf of

Biberach (1270-1330), who teaches that the spiritual senses are activated and perfected by the touch and taste of the Eucharist. A third is Bonaventure (1217-1274). He, like his fellow schoolman Albert the Great, ranks taste and touch the highest of the spiritual senses but, unlike him, he hints at the kind of integral, nondualist concept of spiritual sensation we have seen in Bernard's and Hadewijch's writings.

Ruusbroec, Rudolf, and Bonaventure explicitly state that the spiritual senses of taste and/or touch best attain to God. One reason why is that they, like Bernard and Hadewijch, want to articulate a nonintellectual, immediate relation to God; that is, a union fully informed by piety—by the thoughts and actions of the whole person, not ideas or talk that show no necessary connection to the practice of a virtuous and holy life. They, too, could find warrants to use language of taste and touch to articulate union with God in the idea that we must "experience" such union in the model of Christ and on the basis of our likeness to God as whole people, soul and body.

On the other hand, these men understand the spiritual senses in terms of faculty psychology. They assign the spiritual senses to the *affectus*, the faculty of love and desire—in fact, and more precisely, they identify the *affectus* specifically with the spiritual senses of touch and/or taste. Partly inspired (as was Albert the Great) by the ideas of Thomas Gallus (c. 1190-1246),[2] they combine the idea that the *affectus* is a faculty of knowledge with Dionysian epistemology and present *affectus* as the faculty that allows us to know God in union, when intellect is "blinded" and "dark." They use language of touch and taste to articulate how *affectus* knows: immediately, by direct contact or "experience." This allows them to maintain the idea that the spiritual senses allow us to know God, without associating them with intellect that "sees" or with intellectual knowledge. They therefore invert the traditional hierarchy of senses; they demote sight (when speaking of union with God), retaining it only with a negative meaning to articulate the unknowability of God's transcendence. For although the intellect sees the divine light, it cannot attain to or establish an immediate relation with him; ultimately it fails and sees only darkness. Touch and taste, on the other hand, they present as the spiritual senses best able to attain to God, because they associate touch and taste with knowledge acquired both from bodily contact and from the "experience" of love.

Although Ruusbroec, Rudolf, and Bonaventure use language of taste and touch to discuss union, they shy away from its implications of immediacy and reciprocity. They (like Augustine and the scholastics we looked at earlier) divorce the spiritual senses from the human person by binding them closely to grace, to the point of identifying them with grace. Ruusbroec, for example, emphasizes this connection or identification with grace because he explicitly recognizes the idea that we have an innate something that allows us an immediate relation to God in the manner of sensation. Yet he does so while himself discussing how God and the human person touch, and he has to work hard to evade the implication of reciprocity.

Ruusbroec, Rudolf of Biberach, and Bonaventure use sensory language that strongly suggests a concept of spiritual sensation that seems akin to that implied by Bernard and Hadewijch. On the other hand, they insist on a distinction between the embodied human person and its spiritual senses, which are a divine gift and not innate; in that respect their concepts of the spiritual senses are more akin to the dualist versions of Origen, Augustine, and the scholastics I discussed earlier.

BONAVENTURE

Bonaventure (1217-1274) had more to say about the concept of the spiritual senses than did anyone before him save Origen. He knew about the idea from various sources, including his teacher Alexander of Hales, Anthony of Padua (1191-1231), Thomas Gallus, and Bernard of Clairvaux. He also almost certainly knew the discussion in Origen's *Commentary on the Song of Songs.* Like Origen and Albert the Great (and, in his way, Bernard), Bonaventure develops the concept of the spiritual senses specifically as a mystical doctrine that has to do with how we know God in the highest stages of our journey to union with him. In most respects he echoes the dualist, "Origenist" concept; however, he incorporates ideas and language that challenge that concept.

Bonaventure discusses the spiritual senses in several places, all in connection with grace.[3] That is, he does not discuss them as powers of the soul, but specifies that they are acts that follow on the gifts of grace.[4] In this Bonaventure, like the other scholastics we have glanced at, presents the spiritual senses as different in kind from the corporeal senses in order to underline the gap between spirit and body and the gulf between us and God, which can be spanned only by God. He does *not* say that grace in any way perfects our corporeal senses. Those are intrinsic to the human person and unseverable from their objects in the physical world. The spiritual senses are something else added on by God; they are more of him, more of spirit, than of the human person. They can attain to and know high spiritual things because they are themselves of the spiritual realm.

Bonaventure says clearly enough that the spiritual senses are not human powers, but he is less clear about their object(s) and their place in the human journey to God, and his ambiguity on these issues does in fact associate the spiritual senses with the powers of the soul. According to Bonaventure, our journey to God is also a journey within the self and at the same time a shift from lower contemplation of created things to higher contemplation of spiritual things.[5] In the *Itinerarium,* he divides this journey into three broad movements. In the first, the "purgative way," the soul contemplates the image of God in creatures, by means of sense and imagination, which are its lower and more corporeal powers. In the second, the "illuminative way," the soul contemplates the image of God within the self, by means of the more spiritual faculties of reason and intellect (*ratio, intellectus*). In the third, the "unitive way," the soul transcends itself and contemplates God himself, by means of its highest, properly spiritual aspects, intelligence (*intelligentia*) and *apex affectus*.

That is a very bare sketch of a complex itinerary, but I make it to show that, as he assigns the spiritual senses an object and a particular stage in the journey to God, Bonaventure assigns them a place within the hierarchical structure of the human person. What is interesting and significant is that he assigns two different places. The two sides of his presentation have been well captured by Karl Rahner and Hans Urs von Balthasar, each of whom emphasizes one aspect at the expense of the other.[6] According to Rahner, Bonaventure suggests that the spiritual senses are active in ecstasy, in particular when the highest part of the soul, the *apex affectus*, "touches" God in the "darkness" of unknowing. On this basis Rahner argues that, first, the spiritual senses have as their object God in his transcendence and, second, that they are sharply distinct from the corporeal senses and the lower powers of the soul. According to von Balthasar, on the other hand, Bonaventure teaches that the spiritual senses are active not within ecstatic union but *before* it. Therefore, the object of the spiritual senses is never God in his transcendence but always the Word, especially the incarnate, fully human Christ who mediates divine transcendence to humanity. So according to von Balthasar, Bonaventure teaches not that the spiritual senses are sharply distinct from the corporeal senses and the other lower powers of the soul but that the senses are integrated, in the pattern of Christ's simultaneous humanity and divinity.

Stephen Fields has argued that Rahner's and von Balthasar's divergent interpretations can be traced to their differing theological views.[7] However, I think only partly: Rahner and von Balthasar latched on to different sides of a genuine ambiguity in Bonaventure's ideas. On one hand, Bonaventure (like other scholastics of his generation and just before) strongly distinguishes between the spiritual and the bodily senses, and he ultimately assimilates the spiritual senses to the spiritual power of the soul that knows God—which he considers to be the *affectus*, not the intellect. The strong distinction is most evident where he suggests that the spiritual senses operate within ecstasy. That is when the *apex affectus* knows and achieves union with God in his transcendence, and it seems to do so insofar as it is distinct from the other powers of the soul. On the other hand, Bonaventure ranks touch, traditionally the most somatic and "base" sense, as highest among the spiritual senses and as the most fitting and active in union with God, where its object is the Word incarnate. As von Balthasar argues, Bonaventure seems to elevate the more immediate and somatic senses to emphasize the christological character of union with God. The goal is to follow Francis, who perfectly imitated the human Christ and who, when he ultimately received the stigmata, was touched directly by God in the most palpable and bodily way. By imitating Francis, a person imitates Christ and achieves union with God in the pattern of the Word incarnate. In this respect Bonaventure blurs the dualism of the "Origenist" concept of spiritual senses. Similarly, he writes that the spiritual senses are active in the stages of purgation and illumination, *before* ecstasy, which implies that they involve the lower *and* the higher powers of the soul—that is, not just intellect and will, but also sense and imagination. This placement of the spiritual senses aligns anthropology with Christology: the *whole* person, spiritual and corporeal aspects, contemplates and knows the incarnate Word, God and man, who perfectly encompasses both the spiritual and the created realm.[8]

In his *Commentary on the Sentences,* Bonaventure specifies that spiritual senses are "the use of interior grace with respect to *God himself*" (*III Sent.*, dist. 13, dub. 1; *Opera,* 291b). In the *Breviloquium,* he presents them as "the final ramifications of grace,"[9] and suggests they are active in ecstatic union with the Father. On the other hand, he explicitly links them to the incarnate Word, which implies they are active well before ecstatic union. He introduces them at the end of a chapter on grace, where he discusses how grace progressively transforms the soul, producing habits (*habitus*) that allow the soul to journey to God (*Brev.* 5.6). The ultimate habits are the beatitudes; the highest of these is wisdom, which "disposes one to peace." Peace, in turn, produces spiritual delight and contemplation, in which "the gazes and embraces of groom and bride . . . are possessed according to the spiritual senses." Then a person sees the beauty of Christ, hears the supreme harmony, tastes the greatest sweetness, smells the greatest fragrance, and touches the Word incarnate—or maybe God the Father. Bonaventure writes that "the greatest sweetness is gathered together [*astringitur*] by reason of the incarnate Word, living among us bodily and restoring himself to us palpably, kissably, embraceably through most ardent love, which through ecstasy and rapture causes our mind to "'pass over from this world to the Father'" [Jn 13:1] (*Brev.* 5.5; *Opera* 5:259b).[10] The soul touches the Word, and the Word then lofts the soul into ecstatic union with the Father. The object of the spiritual senses in general is the Word, but spiritual touch also has a special object, God the Father, and therefore a special activity in ecstasy.

Immediately after this passages, Bonaventure implies that the spiritual senses in general are active in ecstasy, and does so in a way that allows us to grasp their relation to the whole person. He writes that "the spiritual senses concern mental perceptions about the contemplation of truth"; this contemplation culminates in "wisdom, the excessive or ecstatic knowledge, which begins in this life and is completed in everlasting glory" (*Brev.* 5.6; *Opera* 5:260a).[11] Key here are the phrases "mental perceptions" [perceptiones mentales] and "excessive or ecstatic knowledge" [notitiam excessivam]. They appear to refer to the highest stages of union with God the Father, when (you will recall from my earlier sketch of the soul's mystical itinerary) the *apex affectus* outstrips all the lower powers of the soul and transcends itself in ecstatic union with God. These phrases therefore imply that the spiritual senses operate at this highest stage of union, and are closely associated with *apex affectus,* which knows and is one with God as other powers of the soul cannot, and which does so largely independent of them.

The same ambiguity shows up in the *Itinerarium:* Bonaventure specifies that the object of the spiritual senses is the mediating Word but also suggests it is God in his transcendence; and again he associates the spiritual senses with two different "locations" in the hierarchical structure of the human person. He introduces the spiritual senses at the ultimate point of the second, illuminative way (which is the way of introspection, when the person enters into themself by way of Christ the mediator), at the end, just before the unitive way and contemplation proper—that is, before the soul directs itself to God himself and can enjoy ecstasy (*Itin.* 4.2). (At this point "the soul is disposed to mental elevation," but does not enjoy it yet; cf. *Itin.* 1.4.) Therefore, the object of the spiritual senses is the Word in Christ, not God. Indeed, Bonaventure writes that the spiritual senses, when restored and activated by the theological virtues, apprehend aspects of the Word. Faith allows a person to hear the lessons of Christ, and see the splendor of his light; hope allows a person to smell the spirated Word; and "when he embraces the incarnate Word in charity, as one receiving from Him delight and passing into Him through ecstatic love, he recovers taste and touch" (*Itin.* 4.3; *Opera* 5:306b). [Dum *caritate* complectitur Verbum *incarnatum,* ut suscipiens ab ipso delecationem et ut transiens in illud per ecstaticum amorem, recuperat *gustum* et *tactum.*] Charity allows a person to touch and taste the incarnate Word—but this Bonaventure calls "ecstatic love." So although Bonaventure specifies that the object of the spiritual senses is the Word incarnate, he also suggests that the senses of taste and touch remain active *within* ecstasy.

This agrees with his treatments elsewhere. Probably influenced by Thomas Gallus, and echoing his fellow Franciscan Giles of Assisi (1190-1262), in two passages Bonaventure correlates stages of contemplation with the hierarchy of the spiritual senses, inverted from the traditional order: from sight, at the bottom, to touch and taste, which characterize the summit of contemplation, rest in God (*requies*).[12] More explicitly still, in his *Commentary on the Sentences* Bonaventure writes that "touch . . . is more perfect and more spiritual than the other senses because it unites the most to him who is the highest spirit. This is why it is said in I Cor 6: 'Who adheres to God is one spirit with him'" (*III Sent.* dist. 13, dub. 1; *Opera* 3:292a). [. . . {T}actus, qui est perfectior inter omnes sensus est spiritualior propter hoc, quod maxime unit ei qui est summus spiritus; propter quod dicitur primae ad Corinthios sexto: *Qui adhaeret Deo unus spiritus est.*] The highest spirit cannot be the Word in Christ: he is the mediator precisely because he is both spirit and flesh. Touch is the most spiritual of the spiritual senses because it can be most perfectly one with spirit *per se.*

The paradox of ecstatic union with God is attained and known by the *apex affectus,* the highest point or aspect of the soul as perfected by grace. The next-highest powers of the soul, *intellectus* and *intelligentia,* cannot grasp God in his transcendence. The *apex affectus,* when perfected by grace, can contemplate not just the fully divine and fully human Christ, not just the highest principle, but the hidden things of God that exceed human thought (*Itin.* 7.1).[13] Therefore, when Bonaventure writes that the spiritual sense of touch best grasps and attains God in union, he associates it closely with the *apex affectus* and suggests that it, like the *apex affectus,* is active in ecstasy. There is no ghost of the corporeal sense here; the distinction between the corporeal and spiritual senses is clear.

On the other hand, Bonaventure also suggests that the spiritual senses are active *before* ecstasy and that their object is not the transcendent God but Jesus Christ the Word incarnate. In this respect he identifies them not with the highest and most spiritual aspect of the soul but with all the soul's powers, including the corporeal senses. This suggests that the human person has a single, integrated sensorium and blurs the distinction between body and spirit in the human person. The most telling clue is that Bonaventure specifically ranks touch as the highest of the spiritual senses and the most fitting and active in union with God. Von Balthasar emphasizes this point: he argues that Bonaventure ranks the spiritual sense touch highest not in service to an apophatic theology of ecstatic union with transcendence but in service to a kataphatic Christological theology of union with the suffering God-man. Touch refers to the immediate contact of bodies. It is somatic, and therefore it is fitting for human encounter with the Word incarnate, who is integrally spirit and body.

Bonaventure himself alludes to the reason, citing Bernard of Clairvaux: touch is the sense appropriate to union because it refers to close contact (see the quotation above, pp. 106-7). When he refers to the sense of touch, he seems to reinforce his insistence that we know God through the Word incarnate by means of spiritual senses that *do* retain a connection to the body and the lower powers of the soul, in the pattern of the God-man. True, he divides the spiritual senses between different faculties of the soul. The spiritual senses of sight and hearing concern the intellect, while the other senses concern the *affectus* (*III Sent.* dist. 13, dub. 1; *Opera* 3:292a). We have already seen that he identifies touch, more precisely, with the activity of the *apex affectus* in union, which is a form of noncognitive "knowing" by means of love, a "loving intuition." This distinction can be interpreted as sharp and absolute or not. Bonaventure opposes this affective "knowledge" to the discursive knowledge gained by intellect, and he also binds them together, because he implies that affective "intuition" completes and perfects intellectual knowledge. Likewise for the spiritual senses, because he divides the spiritual senses between the faculties of affectus and intellect.

In the *Itinerarium* there is a small but significant piece of evidence for Bonaventure's ambiguous concept of the spiritual senses. At the beginning, Bonaventure speaks of the activity of *apex mentis* in the highest state of union (*Itin.* 1.6; *Opera* 5:297b: ". . . sex sunt gradus potentiarum animae . . . scilicet sensus, imaginatio, ratio, intellectus, intelligentia et *apex mentis* seu synderesis scintilla."). But when he comes to actually discuss this state, he writes *apex affectus*—perhaps because *apex mentis* implies that the soul knows God intellectually, whereas *apex affectus* refers the *affectus,* which "knows" by means of love and "experience." In *excessus mentis, apex affectus* passes over to God and is transformed by "ecstatic anointing and totally inflamed affections" (*Itin.* 7.4; *Opera* 5:12b), and intellectual operations are left behind. [In hoc autem transitu . . . oportet quod relinquantur omnes intellectuales operationes, et *apex affectus* totus transferatur et transformetur in Deum.] These statements imply that the affective power of knowing is distinct from and exceeds discursive intellectual knowledge. But Bonaventure equivocates; he also calls ecstasy a "mode of knowing."[14]

I think we must accept this ambiguity, while noting that Bonaventure's overall tendency, supported by several explicit statements, is to allow that both intellect and *affectus* attain to knowledge and to maintain that each contributes to union. He identifies spiritual touch with the activity of the *apex affectus* in union, and so associates it with the knowledge gained by the loving power of the human soul, the *affectus.* Affective knowledge is in one respect opposed to the discursive knowledge gained by intellect, but in another respect it completes and perfects that knowledge. I suggest that Bonaventure assumes that the spiritual sense of touch completes perfects the lower spiritual *and* corporeal senses—that is, a single sensorium—and so integrate their contemplation of the Word incarnate with contemplation of the transcendent Trinity.[15]

We might conclude that Bonaventure's concept of the spiritual senses is basically dualist. Yet his statements that touch and taste are the spiritual senses that best attain God show that he wanted to find positive sensory language—opposed, that is, to negative Dionysian language of blinding darkness and sightless sight—to articulate an immediate union with God that was beyond intellectual knowing but still known. So, like Gallus before him, Bonaventure uses bodily language; he inverts the traditional hierarchy of senses both because of the association between sight and intellect and because the senses of taste and touch operate by immediate contact of bodies, which considers a solution rather than a problem. On the other hand, Bonaventure uses this language to solve a problem that is more epistemological than anthropological. Although he hints at the kind of integral anthropology that Hadewijch embraces, he still presents the spiritual senses as separate and different in kind from the corporeal senses: they are a gift added to the human person by God and not an aspect or an activity of a unified sensorium that is innate to the whole person.

RUDOLF OF BIBERACH

Rudolf of Biberach gives a similar—and similarly two-sided—account of the spiritual senses. He identifies taste as the spiritual sense, or perhaps just sense, that is the most active in union with God. On one hand, he explains that we know God directly in the Eucharist by tasting him—that is, with our corporeal sense of taste; on the other hand, this corporeal taste of the corporeal God activates and perfects the spiritual senses such that we can know and be one with the transcendent God in ecstasy. (The association between the spiritual senses and the Eucharist seems obvious, but Rudolf seems to have been the first to make it explicitly.) He supports this idea and the special status of spiritual taste with a reference to the main supporting biblical verse, Ps. 33:9 ("gustate et videte quoniam suavis est Dominus") and quotes several authorities, the most important of whom is Thomas Gallus. Like Gallus and Bonaventure before him, Rudolf inverts the traditional hierarchy of senses because of the ancient identification of sight with intellectual knowing.[16] Rudolf strongly emphasizes that a person cannot know God by means of intellect but only by means of the *affectus,* which can attain an immediate relation to God in union when intellect is baffled and "darkened." In this respect, again like Bonaventure, Rudolf wants to identify the spiritual senses closely with "spiritual" knowing—nonbodily, even non*human* knowledge of God in his transcendence. However, the corollary is, again, that the spiritual senses are not intrinsic to the human person but an aspect of grace. In this respect Rudolf, like Bonaventure, distinguishes the spiritual senses from the corporeal senses and the body. He associates them very closely with grace, which is of God, and with the *affectus* and the will, the powers of love and desire, which have the closest affinity for and likeness to the God who is Love, and which can attain to an immediate union with God exactly because they are distinct from the other, less spiritual powers of the soul.

Rudolf's major work is *The Seven Roads of Eternity* (which circulated under Bonaventure's name), which is a compendium of quotations that outlines an itinerary to mystical union with God.[17] He discusses the spiritual senses in the sixth chapter (*Iter* 6) which, although not the end of the mystical itinerary, deals with the deepest and most immediate knowledge of God, "experiential foretaste" (experimentalis praegustatio). The previous three chapters (*Itinera* 3-5) mainly focus on God's incomprehensibility. Rudolf emphasizes that in union a person knows God in loving affect (*charitativa affectio*), especially in the degrees of seraphic love, (*amor seraphicus*) (*Sept. itin.* 4; on *amor seraphicus,* see 4.5 art. 3), and ultimately in *excessus mentis,* which, he explains, is mystical darkness (*caligo mystica*) and ignorance (*ignorantia*) above intellect (*superintellectualis*) (*Sept. itin.* 5.6). His examination of the spiritual senses caps these chapters with a discussion of the *positive* knowledge that perfects and surmounts knowledge-in-ignorance: namely, "experience" or "taste." He thereby brings to a crescendo his continuing and overall point: that though we can know God by means of our rational faculty, the *intellectus,* that knowledge is perfected and consummated in "experience" by means of the *affectus* (more precisely, *affectio principalis*).[18] "To attain to God is always to seek him through desire, to find him through cognition, and to touch him through taste"; these "are principally acts and grades of the *affectus* and the will" (*Sept. itin.* 6.6; quoting Hugh of St. Victor, *The Mystical Ark* 4.4). [Ad Deum attingere, est semper per desiderium eum quarere, et per congitionem invenire, et per gustum tangere. . . . Hi autem actus et gradus sunt principaliter ipsius affectus et voluntatis.] Rudolf closely associates *affectus* with the will, and stresses it as the cognitive complement to the *intellectus.* Intellect enables the human person to know God rationally and discursively, but intellect does not attain to union; *affectus* does and allows a person to know God immediately, nondiscursively, and experientially. This experiential knowledge is gained through love, the proper act of *affectus* and the will.

The spiritual senses, especially the sense of taste, are another such act. "The will, not the intellect, possesses the act of tasting, because taste is in charity, as Richard [of St. Victor] says, and charity or love is an act of the will" (*Sept. itin.* 6.6). [Actum gustandi non habet intellectus, sed voluntas; quia gustus est in charitatis, sicut dicit Richardus: sed charitas, vel dilectio, est actus voluntatis.][19] Rudolf's main or at least most evident warrant is familiar from Bernard of Clairvaux's *Sermons on the Song of Songs*: the equation between "experience" and "taste." A person "knows" experientially by means of the *affectus,* which is to say a person tastes by means of the *affectus.* At the summit of contemplative life, at the point of immediate union with God, Rudolf identifies the *affectus* as the "organ" of spiritual taste.

Therefore, Rudolf unequivocally asserts what Bonaventure had only suggested—that the spiritual senses are active in immediate union. In Rudolf's presentation the spiritual sense of taste *is* the activity and/or bond of immediate union. And the implication is the same, only stronger: that, in union, the spiritual senses (taste, at least) operate independent of the rest of the person. He clearly specifies that the spiritual senses are distinct from the corporeal senses (*Sept. itin.* 6.4, quoting Bernard and Gregory the Great), and (like Bonaventure), associates them closely with the highest point of the soul, which achieves union with and knowledge of God precisely insofar as it differs from and outstrips the other faculties of the soul and the rest of the human person.

On the other hand, Rudolf also suggests that the spiritual senses are innate to the human person. Like Bonaventure, he links the spiritual senses closely to grace. The spiritual senses "sleep" beneath the corporeal senses, and can be awakened only by means of grace (*Sept. itin.* 6.5, in quotes from Gregory the Great and Hugh of St. Victor). But he is not as careful as Bonaventure to specify that they are *acts* of grace in us, not innate or new powers. Rather, Rudolf retains the implication that the spiritual senses are an innate component of the person, usually drugged and darkened by sin, but latent and only needing a divine stimulus. The stimulus is Christ; in particular, Christ as present in the Eucharist (*Sept. itin.* 6.5).[20] The Eucharist activates the spiritual senses, beginning at the low end with sight and ending with taste, the loftiest sense, which can "experience" God in the bread and wine.[21] The Eucharist activates the spiritual senses, especially taste, precisely because it is, like Christ, simultaneously spiritual and material. The bread and wine are truly the body of Christ, they make God immediately present, and they allow us immediate knowledge of God's hiddenness. Physical eating of the bread and wine "warms and encourages" the corporeal senses; they in turn stimulate the intellect and especially the will, which in its turn tastes Christ (*Sept. itin.* 6.5). Physical tasting and spiritual tasting are not really distinct but are aspects of each other. "In the sacrament Jesus is true food, and truly tastable, and through his corporeal taste one is extended to his divine taste, because he is the way, according to his humanity, to his divinity." Eating Christ in the Eucharist activates the corporeal sense of taste and the intellect and the will, "and thus by spiritual chewing of the body of Christ one is extended to the taste of hidden divinity" (*Sept. itin.* 6.5). [. . . Per gustum corporis ejus pervenitur ad gustum divinitatis ejus, quia ipsa est via, secundum humanitatem, ad ejus divinitatem, . . . et sic per spiritualem manducationem corporis Christi, {voluntas} ad gustum absconditae divinitatis pertingatur.] The Eucharist is the proper object of the spiritual senses because it shares and conforms to the structure of the human senses—and the human being: spirit and body fundamentally one. (Note the contrast between Rudolf's liturgical conception and Origen's exegetical one; recall that, according to Origen, the spiritual senses are activated by and have as their object the anthropomorphic language of the Bible.)

Rudolf again implies that the spiritual senses are continuous with the corporeal senses when he teaches that contemplative union is only completed in action. After the chapter on the spiritual senses is another culminating chapter on "godly work" (*deiformis operatio;* another possible translation is "work [in]formed by God"), which is the practice of virtue in the world. Rudolf, following Cistercian and Victorine teaching, emphasizes that such work is necessary to complete the mystical life. Although he clearly presents "experiential foretaste" and the spiritual senses as the apex of union properly speaking, he places the "godly work" of virtue in the seventh and final position to teach that such union is only finalized and perfected when it extends to the *whole* person as they live and act in the world. Rudolf cites various authorities to this effect and culminates with a quotation from William of St. Thierry: as by interior works of contemplation a person has a taste of eternal life, so by exterior works they are made the likeness or image of eternal life; thus the whole body serves God and is made harmonious to him, "and truly man and God are one heart and one soul" (*Sept. itin.* 7.6, citing William of St. Thierry, *De natura et dignitate amoris* 20, attributed to Bernard of Clairvaux). The pattern and model is the same one we have seen throughout: Christ, doing his divine work in the world as an embodied human being. Thus Rudolf thinks that, as the Eucharist manifests the dual nature of Christ, so it perfects our dual nature, bringing the whole to union ("experiential foretaste") with God.

Rudolf's ideas about the spiritual senses echo the ambiguity of Bonaventure's account. Rudolf is at pains both to stress the distinctiveness of affective "experience" (its high and discrete status in the hierarchy of human knowledge and of the human being) and to emphasize its continuity with the "lowest" levels of knowledge and person—namely, the sensorial and corporeal knowing that characterizes taste and touch. He clearly conceives of this continuity as a fundamentally Christological integration of spirit and body: that is why the Eucharist is the proper object of the spiritual senses. In this respect Rudolf suggests the unified sensorium that is implied by Bernard, Hadewijch, and Bonaventure. But this integral, Christological concept of spiritual sensation interferes with Rudolf's more dualist conception that "affective experience" can attain to God precisely insofar as it is distinct from other modes of knowing and other powers of the soul.

JAN RUUSBROEC

Rudolf, like Bonaventure, explicitly links the spiritual senses to grace, though he is less careful to reject the idea that they are innate powers of the soul. We have seen that the connection to grace both supports the notion that the sensorium is one (either darkened by sin or redirected and perfected by grace) and threatens to undermine that notion; as acts of grace, rather than powers or faculties, spiritual senses would seem to be secondary effects of an intrinsically divine gift, something added to the human person. This tension is especially prominent in the writings of Jan Ruusbroec. Although Ruusbroec explicitly discusses the doctrine of the spiritual senses as such only a few times, he uses sensory language extensively to articulate union with God and discusses what that language signifies. Most notable is his language of touch (to which I shall confine myself here),[22] which reveals the same basic ambiguity we have seen in

Bonaventure's and Rudolf's concepts of the spiritual senses: a more dualist, "Origenist" concept of spiritual senses partially integrated with a less dualist, more "integral" concept of spiritual sensation.

Where Bonaventure and Rudolf of Biberach seem to integrate the two concepts by identifying the highest spiritual sense (touch and/or taste) with *affectus,* the "faculty" of nonintellectual knowing, Ruusbroec seems to integrate them by identifying spiritual touch with a trinitarian bond between us and God—a bond that is both entirely God and spiritual, and inescapably corporeal, because integral to the *whole* person. This is quite similar to Hadewijch's ideas. We saw that she fully exploited the idea of reciprocity implied by language of touch, which she used to articulate the fundamental equality of *minne* in us and God, and the idea that *minne* was not merely spiritual, but intrinsic to the whole person, soul and body. Ruusbroec may have derived these ideas from Hadewijch. In any case, he uses language of touch for the same reason, to articulate the immediacy of union with God as a kind of reciprocity—but only a kind. In contrast to Hadewijch, he explicitly denies that union is fully reciprocal, not least because he is concerned to refute contemporaries who taught that union with God could be a fully reciprocal relation of equals that divinized a person and raised them beyond all worldly virtues and constraints.[23] In Ruusbroec's formulation, touch is active in God, passive in us: God touches, we are touched.

In other words, Ruusbroec, like Bonaventure, assimilates spiritual touch to grace; in particular, to the grace and gifts that flow into the person at the penultimate stage of the life of union.[24] In *The Spiritual Espousals*, he divides the life of union into three broad stages or aspects: the active life, which is union with intermediary; the interior life of inner yearning, which initiates union with*out* intermediary (*eenicheit sonder middel*); and the contemplative life, in which union without intermediary is perfected by union without difference (*eenicheit sonder anderheit*, *sonder differencie*). He introduces language of touch in the second stage in order to articulate unity of spirit (*eenicheit des gheests*),[25] which is the apex of the interior life and the threshold of contemplative life and union without difference. Ruusbroec identifies this touch with God's action in us—in the "spirit," (*ghedachte*, *gheest*), the highest part of the soul. To the extent that it is a feature of the human person, the spiritual sense of touch seems entirely passive, a mere receiver: "The creature undergoes and suffers this touch. . . . and here no one works but God alone, . . . it is a divine activity, and the origin and incursion of all grace and of all gifts, and the last intermediary between God and the creature" (*Espousals* b 1477-87; *Werken,* 197; translation Rolfson, slightly altered). [Ende dit gherinen doeghet ende lijdet de creatuere, . . . Ende hier en werket niemen dan god alleene. . . . Want dit es een godlijc werc ende oersprinc ende inval alre gracien ende alre gaven, ende dat laste middle tuschen gode ende de creature.] If God's touch (*gods* *gherinen*) is "added on" to the human person, and is then the "spiritual sense" of touch, then it—and by extension, the "spiritual senses" as a group—seems distinct from the human person.[26]

Ruusbroec is usually careful to write that God touches us, not the reverse.[27] But not always. He also writes that we touch God—more precisely, our spirit does:

> In [the] storm of love, two spirits contend: the Spirit of God and our spirit. God, through the Holy Spirit, inclines Himself towards us, and thereby we are touched in love. And by God's operation and the faculty of loving, our spirit presses into and inclines itself towards God, and thereby God is touched. From these two, there arises the strife of love: . . . [in which] each spirit is most wounded by love. These two spirits, that is, our spirit and God's spirit, flash and shine each into the other, and each shows the other its face. This makes each spirit continually crave for the other with love. Each demands of the other that which he is, and each offers and invites the other to that which he is. This makes the lovers flow away [into each other]. . . . This flowing out and flowing back cause the fountain of love to overflow. Thus God's touch and our love's craving become one single love. (*Espousals* b 1558-63; *Werken*, 200; translation Rolfson, slightly altered)

> [In desen storme van minnen striden twee gheeste, die gheest gods ende onze gheest. God, overmids den heilighen gheest, neyghet hem in ons, ende hier af werde wij in minnen gherenen. Ende onse gheest, overmids goods werc ende de minnende cracht, druct ende neyghet hem in gode, ende hier af wert god gherenen. Van desen tween ontspringhet der minnen strijt: . . . {in dat} elc gheest van minnen <meest> ghewont. Dese twee gheeste, dat es onse gheest ende gods gheest, blicken ende lichten die een inden anderen, e{l}nde elc toent anderen sijn aenschijn. Dit doet ee<n> paerlijcke die gheeste met minnen den eenen inden anderen crighen. elc eyschet den anderen dat hi es, ende elc beidet ende nodet anderen dat hi es. Dit doet de minnende vervlieten. . . . Dit voeyen ende dit wedervloeyen doet overvloeyen die fonteyne der minnen. Aldus wert gods gherinen ende onser minnen crighen eene eenvoldighe minne.]

Spiritual touch is a living act of love, and the human person *is* active in it. Ruusbroec seems to appeal to touch precisely because it suggests a two-way, reciprocal relation. Here, as elsewhere, he maintains the Augustinian formulation of grace, and has God touching first. Nevertheless, the person does touch *back*; and if this is not a reciprocal relation of equals, it is a relation of matched and equal acts. That is Ruusbroec's point (as it is Hadewijch's). Both our and God's spirit crave, demand, offer, and invite, such that they flow each into the other. Each craves and each only partly attains; we are partially thwarted, but so is God. Thus the relation is always dynamic, never resting. "The living union that we feel with God is active, and is always renewed between us and God. . . . we mutually kiss and touch. . . . [and] feel that we are both touching and being touched, loving and being loved, always being renewed and returning to ourselves, going and coming like the lightning of the heavens" (*Mirror* 4.C.b; *Werken,* 207). [De levende eenighe die wi met Gode ghevoelen, die es werkelec, ende vernuwet altoes tusschen ens ende Gode. . . . wi underlinghe kussen ende gherinen. . . . ghevoelen wi dat wi gherinen, ende gherinen werden; minnen, ende ghemindt werden, ende altoes vernuwen ende wederkeeren in ons-selven, gaen ende comen also de blixene des hemels.] (Ruusbroec also, like Bernard, and almost in Hadewijch's words, invokes the motif of mutual eating, which strongly suggests that he does want to allow for reciprocity.[28])

It seems clear that Ruusbroec specifies that God touches first, then we touch back, because he wants to preserve the distinction and inequality between God and the person. But one of his major points is that the three basic aspects of union are simultaneous, not sequential. In this life, as in heaven, union is contemplative and also fully purgative and illuminative. The higher stages do not eclipse or supersede the earlier, but complement and deepen them.[29] Ruusbroec uses language of touch to articulate this point because to touch and to be touched are simultaneous, not sequential. In this, his usage is very much like Hadewijch's, and in fact has very similar theological roots. Ruusbroec's theology and anthropology of union, like Hadewijch's, is rooted in a dynamic conception of the Trinity. To achieve union with God is to partake of the inner dynamism of the Trinity, ever sufficient to itself and ever flowing out to creation. Ruusbroec articulates this trinitarian union with language of touch. Like Hadewijch, Ruusbroec identifies the source of spiritual touch in the fecundity (*vruchtbaerheit*) and richness (*rijcheit*) of God (in many places: e.g., *Espousals* b 1475; *Werken,* 197). These terms indicate the inner life of the Trinity: the dynamic relations between the three persons, between them and creation, and between them and the underlying unity from which they proceed (*Espousals* b 1489; *Werken,* 197). The touch of the Trinity is the source of touch in the human person—that is, the touch that brings us to union. Touch is grace that flows out into the innermost depths of our spirit (*innichste sijns gheests*) (*Espousals* b 1502), its unity (*eenicheit*) (*Espousals* b 107ff, b 1496, b 1626ff; *Werken,* 147, 198, 202). In union of spirit, that innermost unity shares the "welling" flow of touch with God, and wellingly flows with him: "It [the grace of God] is situated in the unity of our spirit like a fountain, and it wells in the same unity whence it arises, truly as a living vein welling out of God's richness. . . . And this is the touch I mean" (*Espousals* b 1473-5; *Werken,* 197; translation Rolfson, slightly altered; see also *Der Rike der Gheleven* I.60.2 [quoted in Reypens, 164]). [Ende si houdet hare in eeicheit ons gheests also een fonteyne, ende si walt inder selver eeicheit daer si ontspringhet, rechte also eene levende wallende adere ute eenen levende gronde der rijcheit gods. . . . Ende dit es dat gherinen dat ic meyne.]

Thus Ruusbroec presents union of spirit as an immediate participation in the dynamic inner life of the Trinity, in which the three persons restlessly "flow out" from their unity and then "flow back" into it (see the long passage quoted above, p.114). Like Hadewijch, Ruusbroec calls this dynamism "touch"—an "outflowing" and "inflowing touch" [uutvloeyende, intreckend gherinen]—to articulate both the dynamism of union and its immediacy (*Stone* 726, 728; *Werken,* 32). I say "immediacy" because, although Ruusbroec specifies that spiritual touch is "the last intermediary between us and God," he also says it *is* the immediate union of our spirit with his. It is both source and effect. On the one hand, spiritual touch is produced *by* the fecundity of the Trinity, and so is logically posterior to and distinct from it; on the other hand, it *is* that fecundity, in the Trinity and in us, and it *is* the union between God-as-Trinity and us, and so is logically simultaneous with and indistinct from its "source." *Gherinen* articulates a union in with—and within—the dynamic inward and outward flowing of the Trinity. *Gherinen* is the last intermediary because it paradoxically keeps us from union without difference with the idle emptiness (*ledicheit*) of the *Trinity's* unified ground. The touch of the Trinity is an endless flowing river that drives our desire and draws us in, but it never satiates us and so pushes us back (*Stone* 723-31 and ff.; *Werken,* 32-3). The essential reciprocal (*ongheduerheit*) of touch articulates both reciprocity and the restless, unsatiable craving for rest, its dialectical partner. The union of spirit that is spiritual touch is a mediated union with God in his unity and an immediate participation in the life of the Trinity.[30]

Is spiritual touch of God, or of the person? Is it an act that affects only the "spiritual" part of the person, the part distinct from body, or does it affect and belong to the whole person, body included? The answer seems to be Yes to all. On the one hand, spiritual touch is something "divine" and fully spiritual in the human person; it is a restless spiritual aspect that is (or can be) one with the divine dynamism of the Trinity. It seems to be quite distinct from the lower, less spiritual and more corporeal aspects of the person. We have read that this trinitarian spiritual touch is felt in and emerges from the "innermost depth of our spirit" and its "unity." "Spirit" itself (*ghedacht*, *gheest*) is the highest, spiritual part of the person, distinct from the lower parts of the person, senses and body, with the soul mediating between. Spirit includes intellect and will and, more broadly, the *affectus*, which is the seat of love and the knowledge attained by love.

In other words, Rudolf, like Bonaventure and Rudolf of Biberach, associates spiritual touch and taste with the *affectus* and will, and he does so in order to distinguish the knowledge of God that is gained in union from the intellectual, rational, discursive knowledge possible for the lower self. God's touch actuates a union where the corporeal senses and the intellect have no activity, because the object of union is God's transcendence, which cannot be sensed or known intellectually. This touch "unifies" the higher faculties of spirit (*ghedacht, gheest*) and the virtues, rendering them useless and exceeding them in *unitas spiritus* (*Espousals* b 1479-80; *Werken,* 197). In union of spirit, intellect is blinded while affectus and will are still active and know God by love and by the spiritual senses—especially the "touch of love." Touch is not without reason and intellect but "above" them. When God touches, reason is touched along with the rest of the person, but reason is not sensitive to touch (or does not touch *back*, although Ruusbroec does not put it this way). In fact, touch "silences" the faculties other than *affectus*, where it is received—and returned—independent of the rest of the person. The love of Father and Son do not flow into the whole person, but into us above our will, in the ground of its activity, in the spark of the soul (*Mirror* 3.C.a; *Werken,* 206). "This is the life of love in its supreme activity, above reason and understanding, for here reason can neither give to nor take from love, for our love is touched by divine love" (*Espousals* b 1550-53; *Werken,* 200). [Ende dit es minne leven in haren hoochsten werken, boven redene ende verstaen; want redene en mach hier der minnen gheven noch nemen, want onse minne es van godelijcker minnen gherenen.]

In these respects Ruusbroec's concept of the spiritual senses seems dualist ("Origenist"), because he distinguishes them from the corporeal senses and the rest of the human person. However, Ruusbroec presents the spiritual senses as having a root in each moment of the different stages (or simultaneous aspects) of union. In other respects the spiritual senses seem not strictly spiritual and "divine," but rather of us, of the person—and the *whole* person, spirit, soul and "experienced" body. Spiritual touch is not just of God, but of us. A person feels the touch in their spirit, and they also actively touch back, apparently by means of that same spirit. If in some respects distinct, the spiritual sense of touch seems continuous with corporeal touch and the rest of the human person: Ruusbroec's choice to use language of *touch* itself underlines the continuity with the bodily aspect of the person. The union of spirit that is dynamic spiritual touch manifests itself in the *whole* person and seems to belong fully to the whole human person because the Trinitarian bond with God is necessarily realized in the whole person, including the "experienced" body. Active touching in union necessarily "flows out" not only to God in the spiritual realm, but to neighbor, in embodied work in the world—in what Ruusbroec calls *levende* (active) *lijf*, which is a life of love, virtue, and imitation of Christ.[31]

As I have argued throughout, this simultaneously trinitarian and christological theology of union tends to dissolve any sharp distinction between spiritual and corporeal senses. Even though the spiritual realm is distinct from the created realm of matter, a person cannot know or be one with God only "spiritually," because the spiritual aspect of the human being is not the full self, and it is not the exclusive site of union or salvation. So it makes sense that the immediate contact that gives knowledge must operate both "here" and "there." Hence the concept of a unified, integral sensorium (a "single set of senses") which, if perfected, is able to operate in both realms, and in this life. Without such a sensorium it seems impossible to live a holy life in imitation of Christ, directed to and nourished by immediate relation with God.

Ruusbroec reinforces this integral anthropology when he implies that resurrection will perfect our bodily senses, in the pattern of Jesus Christ. He writes that, in this life, we receive Christ bodily in the sacrament, but we cannot receive Christ as he is now, bodily in heaven, until we are resurrected. Ruusbroec insists that it is impossible to bring eternal things—spiritual and heavenly things—into the temporal and material order. However, "after the Last Day, when we come to heaven in our own glorified body, ... There we will behold his glorious countenance with our bodily eyes and hear his loving, sweet voice with our bodily ears. At this our heart and all our senses will be filled with glory and we will be immersed in him through love and joy, and he in us." Ruusbroec immediately goes on to say that although this is far more than earthly senses can perceive, "all this is the least of the glories of heaven, because it comes from without and *is perceptible by the senses*" (Mirror. 2.B, fourth point; *Werken,* 174; translation Wiseman, my emphasis). [Want na den lasten dagh, alse wi met onsen gloriosen lichame in den hemel comen, . . . ende selen bescouwen sijn gloriose anschijn met onsen lijfleken ooeghen. Ende wi selen hooeren sine minleke soete stemme met onsen uutwendeghen ooeren. Ende hier-ave sal onse herte ende alle onse senne vervult werden met sijnre gloriën, alsoe dat wi versmelten van minnen ende van bliscapen in Heme, ende Hi weder in ons. Ende al es dit de menste glorie des hemels, omme-dat si van buten comt ende senlec es. . . .]

The model here is the resurrected Christ, about whom Ruusbroec says elsewhere that "throughout the days of his life his bodily powers and his power of feeling, his heart and his senses were commanded and invited by the Father to the higher glory and bliss, which he now [i.e., in heaven] tastes sensibly according to these bodily powers" (*Espousals* b 925-28; *Werken,* 177; translation Rolfson, slightly altered) [. . . alle sijn leefdaghe waren sine lijflijke crachte ende sijn ghevoelijcheit, herte ende sen gheeyschet ende ghenoedet vanden vadere, tot diere hogher glorien ende welden diere hi nu ghevoelijcke na den lijflijcken crachten ghesmaect. . . .] Here he implies that resurrection perfects our bodily senses. There seems to be no sharp distinction between this spiritual sensation and perceiving God in heaven with the "glorified" senses.

In some respects Ruusbroec seems to assume that the spiritual and corporeal senses are ends of a continuum, aspects of one unified sensorium. Yet he seems to preserve the dualist conception of spiritual senses, insofar as he reserves full "perception" to union enjoyed by the spiritual and "divine" part or aspect of the person, and insofar as he assimilates spiritual "touch" to grace, which is of God. However, while there is no doubt that Ruusbroec wants to preserve the distinction and inequality between God and the person and to retain the idea that we need God's initiative and help to bridge the gap between him and us, he also wants to talk about union as a relation of perfect immediacy. It is significant that he chooses to call union a "touch." He does not reject such somatic language but accepts the corporeal reference that suggests the continuity between "spiritual" touch and the human sensorium. Although it is not corporeal sensation, it is not clearly distinct from it, either. This ambiguity dovetails with Ruusbroec's insistence that all grades or aspects of union with God, including union "without difference," are achieved by and manifest themselves in the whole, embodied person living in the world, not just in some rarified spiritual aspect.

Epilogue

Ruusbroec was not the last medieval theologian to use language of touch to articulate union with God nor the last to blur the dualism that characterizes one basic concept of the spiritual senses. His writings were read in the fifteenth and sixteenth centuries in Spain (as were widely available scholastic writings on the spiritual senses, such as those of Thomas Aquinas), and probably influenced the sensory rhetoric and what I might call the theology of sensation and perception in the writings of the great masters of Spanish mysticism: Ignatius of Loyola, Teresa of Avila, and John of the Cross. Although John of the Cross, in some respects at least, seems to fall within the dualist and intellectualist camp—which is not surprising, given his knowledge of scholastic theology—Ignatius and Teresa seem to show something of the integral conception of the sensorium and of human person that is present in Ruusbroec's treatment. Detailed studies of these theologian's ideas on this topic already exist; although they need updating, I leave that to others.[1]

Were I to include evidence from a wider range of medieval religious texts, I might be able to write a more heavily plotted history of how medieval theologians use language of taste and touch and of the concept of the spiritual senses. I might then be able to trace a well-defined lineage, arguing that one basic idea ("the spiritual senses") branched into a related but distinct notion ("spiritual sensation") that was then blended back with the first to make a new hybrid. I might be able to trace a detailed genealogy of influence, from the earliest formulations of "the idea" to later versions of "the idea"—a "thing" evolving through minds and years. However, I think that would be inaccurate. It seems to me that plenty of ideas crop up again and again in the writings of people who cannot have known or traced its literary antecedents, or who can have known them only barely. Although I have suggested a lineage in this book, I have only suggested it because I think "*the* doctrine" of the spiritual senses is not a persisting "thing" and is better understood as a possible and recurring answer or type of answer to a persisting question or problem. That is, the answer has no lineage, unless we impose one; but the driving questions have a history. So rather than write a narrative about the "development" of the persisting thing across the centuries, I have compose a group of stories about how certain people used sensory language to address a small set of similar answers to a fundamental question or problem in Christian thought, one that seems all but inevitable given some of the basic ideas that medieval Christians inherited from their ancient teachers: the cosmos is divided into spirit and matter; God is all spirit; we are only partly spirit; only like knows like; we know by means of our senses. That question is: Do we know God and can we be one with God as "perfected" souls or spirit, or as whole people? I have argued that some medieval thinkers thought that the answer is "As whole people," and that their answer shows up sometimes in an explicit doctrine of the spiritual senses, and more often in the sensory language they use and how they use it. I have argued that some theologians embraced language of touch and taste—which they recognized as somatic language—because they thought we can achieve union with God in the body and in this life, and that such union is not segregated, independent and without consequences for that life, but fundamentally one with it. Union with God is no less vivid and immediate than life in the world, and is not, fundamentally, apart from or away from the world.

To make these points, I have argued that theologians' choice of language tells us something about their theological ideas and that, conversely, their theological ideas allow us to understand why they chose to write one way or another way, with one kind of language or another. I have argued that sensory language necessarily refers to the basic bodiliness of the senses and to the relations between bodies that are implied by different senses. Therefore, sensory language inevitably implies something about the structure of the human person. I have also argued here that the highly somatic language we find in medieval mystical texts articulates concepts about God. The cases of Bernard of Clairvaux and Hadewijch illustrate that sensory language can be used as a theological vocabulary that relates the structure of the person to the structure of God, independent of any formal "doctrine" about how such sensory language applies to God and independent of the issues peculiar to biblical exegesis. Bernard and Hadewijch use language of taste and touch to discuss how the self and God achieve—or better, realize—a fundamental unity. Their usage implies a concept of spiritual senses, if we choose to draw it out. More directly, I think that one reason they and others use this language as they do is that it helps them come at a theological conundrum that they are especially concerned about: how can we and how do we know God as embodied selves?

The idea that human beings have spiritual senses addresses central theological questions about the nature of the human person, God, and language, and the relationship between these three things. Some modern theologians have worked to resuscitate versions of it, but I would emphasize that it only works and is only needed within a dualist cosmology and anthropology. To the extent that modern theologians reject medieval dualism, they reject the idea of spiritual senses. It is possible to graft the idea onto the modern theology of "experience," but I think that is an intellectual dead end, because the concept "religious experience" is an unworkable mess (logically speaking; it is of course a powerful idea, rhetorically speaking).[2] However, if we interpret the concept of the spiritual senses as a formalized extension of the less formalized use of sensory language—as a rationalization for a way of talking—then it is still significant now, insofar as it directs our attention to how people make religious claims—in particular, to how people talk about their bodies and about their "experience" when they talk about God. It helps us pay attention to an important aspect of religious rhetoric. That is a topic that deserves plenty of careful attention, because religious talk has always been and still is wielded to inspire, persuade, and compel people in this world. And this world is where we are.

Notes

Notes to Chapter 1

1. Other Christians and philosophers of his era discuss the idea of spiritual senses, if less thoroughly, and there is evidence of Hellenistic antecedents. As far back as Plato we find the idea that humans have "two eyes," one of the body, the other of the soul (*Symposium* 219a; *Sophist* 254a; *Republic* 519b; *Phaedo* 99e). Only Origen and his younger contemporary Plotinus have fully worked-out theories, but in his study of the idea, Karl Rahner notes antecedents and echoes in Philo, *The Martyrdom of Polycarp,* Clement, and Tertullian ("Le début d'une doctrine des cinq sens spirituels chez Origène," *Revue d'ascétique et de mystique* 13 [1932]:112-45; p. 115, n. 8; hereafter cited as Rahner 1932). Henri Crouzel, in his study of Origen, refers us also to Theophilus of Antioch, *A. Autolycos* 1.2 (*Origène et la "connaissance mystique"* [Desclée de Brouwer, 1961], 178; hereafter cited as Crouzel 1961). Plotinus's concept of spiritual senses can be found in *Enneads* 6.7 and has been analyzed by John Dillon, along with some cryptic antecedents in Albinus, Philo, and Gnostic texts ("*Aesthésis Noété*: A Doctrine of Spiritual Senses in Origen and Plotinus," in *Judaica et Hellenica: Hommage à V. Nikiprowetsky,* ed. A. Caquot, et al. (Louvain and Paris: Peeters, 1986), 443-55; hereafter cited as Dillon). On Clement, see Marguerite Harl, "La 'bouche' et le 'coeur' de l'apôtre: deux images bibliques du 'sens divin' de l'homme ('Proverbes' 2,5) chez Origène," in *Forma Futuri: Studi in onore del Cardinal Michele Pellegrino* (Turin: Bottega d'Erasmo, 1975), 17-42; hereafter cited as Harl 1975. I note that in *De natura Deorum* 18.49-19.49, Cicero reports what seems to be an Epicurean version of the concept. I would guess, although I have not been able to confirm it, that there are also Jewish antecedents.
2. The doctrine of the spiritual senses has received a certain amount of attention from modern scholars. The most prominent of these is probably the Catholic theologian Karl Rahner, who analyzed Origen's concept and sketched a history of its medieval Latin descendants in two articles: Rahner 1932 (cited in the

previous note) and "La doctrine des 'sens spirituels' au Moyen Age, en particulier chez saint Bonaventure," *Revue d'ascétique et de mystique* 14 (1933):263-99; hereafter cited as Rahner 1933. Rahner focuses on Origen, the founder, and Bonaventure, whose teaching on the spiritual senses Rahner judges the apex of its medieval development in the West. There are several other studies of Origen's doctrine, which I will cite in the next chapter, but Rahner's account remains the most influential, as does his history of the spiritual senses generally. Rahner built on (and argued with) the earlier work of Auguste Poulain in *Les grâces d'oraision,* 6th ed. (Paris: 1901; translated as *The Graces of Interior Prayer* by Leonora L. Yorke Smith [London: Kegan Paul, 1910; reprint, St. Louis: Herder, n.d.]). At the same time that Rahner was working on his account, Mary Anita Ewer discussed the spiritual senses in *A Survey of Mystical Symbolism* (London: S.P.C.K and New York: Macmillan, 1933), 31-56, 192-95; hereafter cited as Ewer. Later, Hans Urs von Balthasar weighed in, at least partly in direct opposition to Rahner's views: see *The Glory of the Lord: A Theological Aesthetics,* 7 vols., ed. Joseph Fessio, S.J., and John Richies (San Francisco: Ignatius, 1982-1989); hereafter cited as von Balthasar. Von Balthasar offers both general remarks on medieval and later developments (von Balthasar 1:365-80), and a detailed account of Bonaventure's concept (von Balthasar 2:309-26). Another more superficial account is M. Olphe-Galliard, "Les sens spirituels dans l'histoire de la spiritualité," in *Nos sens et Dieu,* ed. Claude Baudoin *et al.* (Desclée de Brouwer, 1954), 179-93. Rahner's history of the doctrine remains the most complete, but it focuses on Latin writings. That imbalance is now rectified by Mariette Canévet's survey article, "Sens spirituel," in *Dictionaire de spiritualité* 14:598-617 (hereafter cited as Canévet), which puts more emphasis on developments among medieval Greek-speaking Christians and among sixteenth-century theologians; and by Bernard Fraigneau-Julien's study, *Les sens spirituels et la vision de Dieu chez Syméon le Nouveau Théologien* (Paris: 1985) (hereafter cited as Fraigneau-Julien), the first half of which examines several early Greek versions of the doctrine. There is a larger number of articles on the teaching of individual theologians. For Latin authors of the period before 1400 (my focus), most studies deal with Bonaventure (I will cite these when I discuss his teaching in Chapter 5). There are major studies of only a few other theologians: Paul-Ludwig Landsberg, "Les sens spirituels chez Saint Augustin," *Dieu Vivant* 11 (1948):81-105; James Walsh, "Guillaume de Saint-Thierry et les sens spirituels," *Revue d'ascétique et de mystique* 35 (1959):27-42; P. Doyère, "Ste Gertrude et les sens spirituels," *Revue d'ascétique et de mystique* 36 (1960):429-48; Gregorio Penco, "La dottrina dei sensi spirituali in Gregorio Magno," *Benedictiana* 17 (1970):161-201 (hereafter cited as Penco) and Margot Schmidt, "Elemente der Schau bei Mechthild von Magdeburg und Mechtild von Hackeborn: Zu Bedeutung der geistlichen Sinne," in *Frauenmystik im Mittelalter,* ed. P. Dinzelbacher and D. Bauer (Ostfildern: Schwabenverlag, 1985), 123-51. I do not cite a number of brief discussions of the spiritual senses in various studies of medieval theologians. Most of these scholars have followed Rahner's lead, confining their attention to explicit discussions of all five spiritual senses and interpreting language that implies spiritual corollaries to any of the canonical five senses to be fragments of the fully elaborated doctrine that Rahner was interested in.

3. This ranking, essentially a moral one, dates back at least to Plato, who implies it repeatedly: e.g., *Timaeus* 47. Aristotle makes it explicit: see *Metaphysics* 980; *Nichomachean Ethics,* 1117; *Dialogues,* 1984; *Sense and Sensibilia,* 437; *Problems,* 949-50; and esp. *On the Soul,* 413-29, where he distinguishes touch and taste as base and "animal" senses, opposed to the other "human" ones.
4. Modern scholars who pay attention to sensory language tend to share this preference, or oversight. Discussions of the spiritual senses frequently deal only with language of hearing and sight—which is, admittedly, the most common (e.g., Landsberg, Doyère, and Schmidt) This is can easily lead to misinterpretations, because it ignores the fact that this restriction is itself a choice, and one significant for both the writer and the modern interpreter. On the other hand, language of sight does seem to be ubiquitous in human languages. We have excellent evidence that this is because *Homo sapiens* have brains organized to gain most information about the world by sight (and hearing) and process that information using the same neural structures in the brain. Recently a crowd of anthropologists and cultural critics have examined "oculocentrism," so-called, as a feature of Western culture, generally with a negative sign attached and with more attention to contemporary antimodernist political polemic than to historical and physiological evidence. For an introduction to this line of talk, see the collection edited by David Howes, *The Varieties of Sensory Experience: A Sourcebook in the Anthropology of the Senses* (Toronto: University of Toronto Press, 1991), esp. his introductory essay, 3-21; two other examples are Stephen Tyler, "The Vision Quest in the West or What the Mind's Eye Sees," *Journal of Anthropological Research* 40 (1984):23-40; and Constance Classen, *Worlds of Sense: Exploring the Senses in History and Across Cultures* (London and New York: Routledge, 1993). Martin Jay has written an analytical history of the "oculist" line of argument in *Downcast Eyes: The Denigration of Vision in Twentieth-Century French Thought* (Berkeley: University of California Press, 1993), which includes a rich bibliography. A promising group of more historical papers was shared at an international conference on *Les cinq sens au moyen age* in October 1999 at the Univeristé de Lausanne-Dorigny; it is promised that these papers will be published in the journal *Micrologus: Nature, Sciences and Medieval Societies.*
5. This observation has been made by John Giles Milhaven with reference to Hadewijch and noting the influence of Bernard: *Hadewijch and Her Sisters* (Albany: SUNY Press, 1993), 15 (e.g.); hereafter cited as Milhaven. But I disagree with much of his analysis, as I will explain in Chapter 4.
6. There are many studies touching on this shift, although the reasons for it remain unsettled. Three clear overviews with bibliographical notes are Ewert Cousins, "The Humanity and The Passion of Christ," in *Christian Spirituality II: High Middle Ages and Reformation,* ed. Jill Raitt, Bernard McGinn and

John Meyendorff (New York: Crossroad, 1988), 375-91 (hereafter cited as Raitt et al.); Richard Keickhefer, "Major Currents in Late Medieval Devotion," in Raitt et al., 75-108; and Caroline Walker Bynum, "Religious Women in the Later Middle Ages," in Raitt et al., 121-39.

7. André Vauchez, *The Laity in the Middle Ages,* ed. and introduced by Daniel E. Bornstien, trans. Margery J. Schneider (Notre Dame, IN: University of Notre Dame Press, 1993), 219-20.
8. The phrase is Caroline Walker Bynum's. She has argued this point in several studies: see *Holy Feast and Holy Fast: The Significance of Food to Medieval Women* (Berkeley and Los Angeles: University of California Press, 1988) (hereafter cited as Bynum 1988); *Fragmentation and Redemption: Essays on Gender and the Human Body in Medieval Europe* (Cambridge, MA: Zone, 1991) (hereafter cited as Bynum 1991); and *The Resurrection of the Body in Western Christianity, 200-1336* (New York: Columbia University Press, 1995) (hereafter cited as Bynum 1995).
9. In some respects my thesis parallels the argument made by Fraigneau-Julien. He distinguishes two "currents" in the way Greek-speaking Christians used sensory language. One describes "intellectual" visions of God using metaphors of light and sight and is represented by Origen and Maximus the Confessor. Another describes more "affective" encounters with God in terms of "presence" or "fullness," which Fraigneau-Julien takes to be metaphors of touch and taste, and is represented by Macarius and Gregory of Nyssa. He suggests that, after Gregory of Nyssa, both these currents were used to describe the highest stages of ascent to God. He finds the ultimate synthesis in the writings of Symeon the New Theologian.
10. This idea has been much discussed in recent years by philosophers and anthropologists interested in metaphor, the philosophy of mind, and cognitive neuroscience. A few who have influenced me are George Lakoff and Mark Johnson, *Metaphors We Live By* (Chicago: University of Chicago Press, 1980); Michael Jackson, "Thinking Through the Body: An Essay on Understanding Metaphor," *Social Analysis, Journal of Cultural and Social Practice* 14 (Dec 1983):127-49; Mark Johnson, *The Body in the Mind: The Bodily Basis of Metaphor, Imagination, and Reason* (Chicago: University of Chicago Press, 1987); F. Varela, E. Thompson, and E. Rosch, *The Embodied Mind* (Cambridge, MA; MIT Press, 1992); and Antonio Damasio, *Descartes' Error: Emotion, Reason and the Human Brain* (New York: Avon Books, 1994). See also the stimulating range of arguments in Sheldon Sacks, Sheldon, ed., *On Metaphor* (Chicago: University of Chicago Press, 1978) and James Fernandez, ed., *Beyond Metaphor: The Theory of Tropes in Anthropology* (Stanford, CA: Stanford University Press, 1991).
11. Hans Jonas, "The Nobility of Sight," in *The Phenomenon of Life: Toward a Philosophical Biology* (Chicago: University of Chicago Press, 1982), 135-56.
12. Albert Deblaere has attempted this with explicit reference to mystical texts, noting that mystical writers appeal to touch to discuss the first contact with

God because it is the "most direct" of the senses: see "Témoinage mystique chrétien," *Studia Missionalia* 26 (1977):127.

13. I discuss these passages in later chapters: Bernard, chap. 3, p. 52; Bonaventure, chap. 5, p. 107; Rudolf of Biberach, chap. 5, p. 111. Bonaventure's statements are discussed by Etienne Gilson in *The Philosophy of St. Bonaventure,* trans. Dom Illtyd Trethowan and Frank J. Sheed (Paterson, NJ: St. Anthony Guild Press, 1965; original French edition, Paris: Vrin, 1954), 420-21; hereafter Gilson 1965. Gilson makes a distinction between knowledge based on "mental representation" and knowledge based on immediate contact.
14. At some point I came across the phrase "rhetoric of sensation" in an article by Perry Miller about the eighteenth-century divine Jonathan Edwards. I lost the reference, but vague credit is better than none at all. Miller too argues that sensory language in religious texts appeals to the body in order to explain and persuade an audience to accept theological concepts.
15. Elliot Wolfson, *Through a Speculum that Shines: Vision and Imagination in Medieval Jewish Mysticism* (Princeton: Princeton University Press, 1994).
16. David Chidester, *Word and Light: Seeing, Hearing and Religious Discourse* (Urbana and Chicago: University of Illinois Press, 1992), esp. 1-30; hereafter cited as Chidester. Chidester analyzes language as a means to discuss the categories symbol, myth and ritual. He links them with the concept "symbolic models," which he defines as "a consistent range of symbolic associations derived from [the distinctive characters of] seeing and hearing" (Chidester, 25). Thus "a visual model tends toward associations of continuity," simultaneity, and detached contemplation of order, while "a verbal model tends to . . . associations of discontinuity," engagement and action (Chidester, 25.). Most powerful are his analyses of how Athanasius and Arius articulated their dispute with different sensory language and clashing symbolic models and of how Augustine develops a subtle rhetoric of sight and hearing to articulate the relation between the self and God (which I will touch on in the next chapter).
17. See also, e.g., Gilson 1965, 416-21, esp. 420; Olphe-Galliard, Walsh, Doyère, Penco, and Fraigneau-Julien; and a recent example is Rosemary Drage Hale, "'Taste and See for God is Sweet': Sensory Perception and Memory in Medieval Christian Mystical Experience," in *Vox Mystica: Essays on Medieval Mysticism,* ed. A.C. Bartlett et al. (Cambridge: Brewer, 1995), 3-14. The same is true for other scholars who only touch on the doctrine in passing: for example, Bernard Bonowitz remarks that "behind the tradition of the spiritual senses, so frequently employed by Bernard [of Clairvaux], stands the assertion that the Divine Object *is* in some way experienced and that these senses are the organs of that experience" ("The Role of Experience in the Spiritual Life," in *La dottrina della vita spirituale nelle opera di San Bernardo di Clairvaux,* Analecta Cisterciensia, 46 [Rome: Editiones Cistercienses, 1990], 323). An exception is Mary Anita Ewer. Her account, although it is brief and cites only a few passages from a narrow selection of sources, is an interesting contrast to the treatments of Poulain, Rahner, and later authors, because she treats the spiritual senses as *language* and notes how references to the senses work as

"symbol" to articulate relations with God: see Ewer, 31-56, 192-95, and 20-30.

18. Stephen Fields has analyzed the dispute in order to highlight the two theologians' differing theological emphases and methods: see "Balthasar and Rahner on the Spiritual Senses," *Theological Studies* 57 (1996):224-41; hereafter cited as Fields. I will touch on his argument in Chapter 5.
19. I cite some examples with specific reference to Bernard and Hadewijch in Chapters 3 and 4. For a concise introduction to the tendency and responses to it, see the Appendix in Bernard McGinn, *The Foundations of Mysticism: Origins to the Fifth Century* (New York: Crossroad, 1991), 265-343 (hereafter cited as McGinn 1991) and idem, *The Flowering Of Mysticism: Men and Women in the New Mysticism, 1200-1350* (Crossroad, 1998), 24-30 (hereafter cited as McGinn 1998). Some samples from the debate are the essays in Stephen Katz, ed., *Mysticism and Philosophical Analysis* (New York: Oxford University Press, 1978) (hereafter cited as Katz); the recent very clear discussion in Mark A. McIntosh, *Mystical Theology: The Integrity of Spirituality and Theology* (Oxford and Malden, MA: Blackwell, 1998), esp. 136-45 (specifically on Rahner, see 91-101), and the excellent essay of Robert Sharf, "Experience," in *Critical Terms for Religious Studies,* ed. Mark C. Taylor (Chicago and London: University of Chicago Press: 1998), 94-116; hereafter cited as Sharf.
20. Indeed, some people would balk at calling Hadewijch's writings "theology" or her a "theologian" and would consider it difficult or pointless to compare her ideas with those of Bernard, much less compare them with the conceptual systems of the university-educated male clerics who are the automatically classed as "theologians." I use the word "theology" broadly to label writing about God and his relation to people and the world, not to select a social group or their favored genres of exposition.
21. A widely-read current example is the collection of Elizabeth Alvilda Petroff, ed., *Medieval Women's Visionary Literature* (Oxford and New York: Oxford University Press, 1987); another is Grace M. Jantzen, *Power, Knowledge and Christian Mysticism* (Cambridge: Cambridge University Press, 1995), hereafter cited as Jantzen; see esp. chap 5 on visions and authority. For a more subtle discussion, see Barbara Newman, "Hildegard of Bingen: Visions and Validation," in *Church History* 54 (1985):163-75; and Frank Willaert, "Hadewijch und Ihr Kreis in den 'Visioenen,'" in Kurt Ruh, ed., *Abendländische Mystik im Mittelalter* (Stuttgart: J.B. Metzlersche, 1986), 368-87, esp. 368-70; hereafter cited as Willaert. For an interesting discussion that confronts this interpretive problem with reference to a text with painted images, see is Jeffrey Hamburger's very interesting book *The Rothschild Canticles: Art and Mysticism in Flanders and the Rhineland circa 1300* (New Haven: Yale, 1990). Hamburger emphasizes, on one hand, that texts and images seem to have been used as "vehicles of mystical elevation," which I think is the interpretation that gives us the most persuasive results; on the other

hand he also assumes they are "simulacra" that *record* experience: see esp. 4-7, 162-67.

22. Denys Turner, in his recent book *The Darkness of God* (Cambridge and New York: Cambridge University Press, 1995), argues that medieval authors who wrote "negative theology" used the basic metaphors of apophatic language (e.g., darkness and interiority) dialectically, specifically to undercut claims about experience of God. Only in the sixteenth century did such language come to be psychologized and interpreted as records of and stimulants to "negative *experience*."
23. I doubt there is any persuading to be done: it seems to me that, on this subject, people are willing to have their ideas confirmed but not denied. Anyone interested in the persuasiveness of a denial could do worse than to read David Hume, "Of Miracles," in *Writings on Religion*, edited by Anthony Flew (La Salle, IL: Open Court Press, 1992), 63-88, and the other writings collected in that book.
24. I do not mean that medieval theological texts give us evidence only for claims about the conscious thoughts of their authors. I work with the premise that writers often imply more and different things than they explicitly state—that the way they put something can directly contradict their explicit statement. I assume further that we can discover some of these unspoken assumptions by carefully analyzing what and how they write. Needless to say, I do still assume the author is an agent with intention, not a ghostly projection of exploded eighteenth-century rationalism. That intention can probably not be pinned down perfectly—who can parse even their own motives?—but it is seems to me silly to deny that people do things with some purpose, and that at least some of that purpose or intention is evident in the results. Some, of course, would argue the point—but apparently to no purpose.
25. Michael A. Sells, *Mystical Languages of Unsaying* (Chicago: University of Chicago Press, 1994). Sells does not entirely sidestep the problems inherent in using such language as evidence for experience. He argues that "apophatic language attempts to evoke in the reader an event that is—in its movement beyond structures of self and other, subject and object—structurally analogous to the event of mystical union." This seems satisfactory for Sells' purpose, which is to "understand the apophatic displacement of the grammatical object as a key moment in a distinctive literary mode . . ." (10). However, although I am persuaded by the way Sells analyzes the actual language of the texts he examines, in order to do so he still relies on the concept "experience"; he simply calls it "event" (or sometimes "meaning event"). He makes an explicit statement that he is avoiding the term "experience," but does not clearly distinguish it from "event." Perhaps by "evoke . . . an event" he means "think" or "produce a state of consciousness," but this is only a guess. Nevertheless, I find Sells' approach much more powerful and persuasive than that of scholars who assume that apophatic language records or transcribes an otherwise ineffable "mystical experience" because he *does* carefully and precisely analyze how that language works, grammatically and rhetorically.

26. For example, Joris Reynaert, who outlines how Hadewijch uses sensory language to articulate theological concepts, also asserts that her somatic language, which includes references to the senses, "records" her somatic experience: *De Beeldspraak van Hadewijch* (Tielt and Bussum: Uitgeverij Lanoo, 1981), 207; hereafter cited as Reynaert 1981. I will discuss this issue more in chapter 4.
27. Some scholars have paid significant attention to sensory language other than language of sight and hearing in medieval texts. The best-known example is Caroline Walker Bynum, who in Bynum 1988 partly deals with sensory language in texts by women and with the question of what this has to do with their "experience" in the world. Another example whose work is directly relevant to my argument here is Joris Reynaert, in Reynaert 1981. I will consider both their arguments in more detail in Chapter 4. Two others whose work is analytically less profound are Wolfgang Riehle, who has described and compared how sensory metaphors are used by *The Middle English Mystics* (London: Routledge and Kegan Paul, 1981), 104-27, without offering much analysis about why; and Marilyn Webster, who has written a literary study of "Mechthild of Magdeburg's Vocabulary of the Senses" (Ph.D. dissertation, University of Massachusetts Amherst, 1996), which is also, unfortunately, largely descriptive.
28. "Objective synaesthesia" or "rhetorical synaesthesia" has at least one other manifestation, which are claims about "synaesthetic experience"—namely, a person's claim that they taste a color or whatnot. Since any such claims are as extraordinary as any biblical miracle, they should be treated as claims only—that is, as rhetorical appeals to the authority of experience—unless there is some corroborating evidence (i.e., some evidence other than the speaker's bare claim).
29. The idea that our senses are five may be too neat, but despite differences in nomenclature, it seems few people think otherwise. On the other hand, the fact that synaesthetic rhetoric ("objective" or "rhetorical" synaesthesia) is so common suggests that people both accept the classification of sensation into five discrete "kinds" *and* consider it inadequate. Joseph M. Williams gives some support to the "givenness" of the senses in his paper "Synaesthetic Adjectives: A Possible Law of Semantic Change" (*Language* 52, no. 2 [1976], 461-78), where he shows a regular pattern to the way lexemes transfer between "sensory modalities."
30. Chidester comes close to this formulation (Chidester, 14-24), although he speaks of the clash of symbols and "symbolic models" that are based on different senses. His analysis is less useful than it might be, because he is happy to use the term"synaesthetic experience" to refer to our perception of events by means of more than one sense at a time (e.g., the mass). By this definition, we almost never leave "synaesthetic experience," and the term is nearly useless.

NOTES TO CHAPTER 2

1. The most significant passages are as follows: *On First Principles* 1.1.7-9; 1.2.7-8; 2.4.3-4; 4.4.10; *Dialogue with Heraclides* 10-12, 16-24; *Against Celsus* 1.48; 2.72; 7.34; *Commentary on the Song of Songs* Prol. 2, par. 9-12; 1.4, par. 24-26; 2.9, par. 12; 3.12; *Homiliae in Lucam* frags. 77, 81 (as designated in Sources Chrétiennes 87); *In Mattheum commentatorium series* 63-64; *Commentarium in Johannem* 10.40; 13.24; *Homiliae in Levticum* 3.3, 31.7; *Homiliae in Ezechiam* 11. I cite Origen's works by book, chapter, and paragraph, according to the following editions and translations. *On First Principles* (hereafter cited as *Principles*): *Traité des principes [De principiis]*, 5 vols., ed.and trans. by Henri Crouzel and M. Simonetti, Sources Chrétiennes 252, 253, 268, 269, 312 (Paris: Cerf, 1978-1984); translations are my own or (if so indicated) slightly adapted from *On First Principles,* trans. by G. W. Butterworth (Gloucester, MA: Peter Smith, 1973); hereafter cited as Butterworth. *Dialogue with Heraclides* (hereafter cited as *Heraclides*): *Entretien d'Origène avec Héraclide,* ed. and trans. (into French) by Jean Scherer, Sources Chrétiennes 67, (Paris: Cerf, 1960); translations are from *Treatise on the Passover and Dialogue with Heraclides,* trans. and annotated by Robert J. Daly, Ancient Christian Writers 54 (New York: Paulist Press, 1992). *Commentary on the Song of Songs* (hereafter cited as *On the Song*): *Commentaire sur le Cantique de Cantiques [Commentarium in Canticum canticorum],* 2 vols, ed.and trans. (into French) by L. Brésard, H. Crouzel, and M. Borret, Sources Chrétiennes 375, 376 (Paris: Cerf, 1991-1992); translations are my own, but I have benefitted from the translation in *Origen: The Song of Songs, Commentary and Homilies,* Ancient Christian Writers 26, trans. with an introduction and notes by R. P. Lawson (New York and Ramsey, NJ: Paulist Press, 1956). *Against Celsus* (hereafter cited as *Celsus*): *Contre Celse.* 5 vols. Ed. and trans. (into French) by M. Borret, Sources Chrétiennes 132, 136, 147, 150, 227 (Paris: Cerf, 1967-1976); translations are from *Against Celsus,* trans. with an introduction and notes by Henry Chadwick (Cambridge: Cambridge University Press, 1953). Chadwick's widely-available English translation of *Celsus* enumerate sections according to the older critical edition of P. Koetschau, *Origenes Werke* (Leipzig, 1899); since this enumeration is also included in the new Sources Chrétiennes edition, I use it also. For other works that I cite only in passing, I do not give full bibliographic information but only standard section designations that will guide a reader to the passage in any published edition or translation.
2. E.g., *Principles* 2.4.3-4; *Celsus* 6.61-62; 2.72; 7.35. This point of departure is emphasized in the following studies: Rahner 1932, 115-16; Crouzel 1961, 262; Henri Crouzel, *Origen,* trans. A. S. Worrall (San Francisco: Harper and Row, 1989), 239 (hereafter Crouzel 1989); Dillon, 443, 449, 455; and Harl 1975, 18, n. 2 (citing Crouzel 1961).
3. Dillon compares the discussion in *Principles* (written c. 230) with the discussions in *Heraclides* (written c. 244-245), *Celsus* (written c. 244-245), and

Comm. ser. Matt. (also written 244-249). I think Dillon's point applies equally well to *On the Song,* which was written c. 245-247, but which he excludes because it survives only in a Latin translation. (I follow the dates given in *Der Kommentar zum Evangelium nach Mattäus,* ed. and trans. with notes by Hermann J. Vogt [Stuttgart: Hiersemann, 1993], Appendix IV, 377-91, which are based on the conclusions of Pierre Nautin, *Origène. Sa vie et son oeuvre* [Paris: 1977].) Rahner also notes that Origen's various accounts are somewhat different and asserts that it was Origen's polemic against Celsus that moved him to develop the doctrine fully (Rahner 1932, 115-17); however, Rahner then smoothly arranges the various accounts into a systematic and harmonious doctrine based on his reading of the later texts.

4. E.g., Rahner 1932; Crouzel 1989; Bernard McGinn, *The Foundations of Mysticism: Origins to the Fifth Century* (New York: Crossroad, 1991), 108-30 (hereafter cited as McGinn 1991); Bernard Fraigneau-Julien, *Les sens spirituels et la vision de Dieu chez Syméon le Nouveau Théologien* (Paris: 1985), 29-24.
5. I do not want to contribute to the endless and misguided wrangle about whether Origen dresses the Bible in Plato or Plato in the Bible. As Jonathan Z. Smith has so eloquently argued in his book *Drudgery Divine* (Chicago: University of Chicago Press, 1990), this kind of opposition between "Hellenistic" and "Hebrew" tells us a great deal about Protestant polemic against Catholic theology, and the reverse, but little about Origen. I find no persuasive evidence that Origen distinguished "philosophy" from his religious beliefs and his practice of biblical interpretation. However, he did write different kinds of arguments. In some texts and passages he draws heavily on the language and concepts of polytheistic philosophers, evidently because he thought that, if he used these terms, the many pagans and Christians who knew and understood them would find his claims comprehensible and persuasive; in other passages he draws more heavily on biblical terminology, for similar reasons. Therefore, I would distinguish not between Origen's more "biblical" and his more "philosophical" concepts but between his different types of arguments and between the rhetorical reasons for those arguments.
6. Origen's principles of exegesis have never lacked attention. For a thorough bibliography through 1982, see Crouzel, *Bibliographie critique d'Origène* (The Hague: Nijhoff, 1971) and *Supplément* (Nijhoff, 1982). The most influential account has been R. P. C. Hanson's *Allegory and Event: A Study of the Sources and Significance of Origen's Interpretation of Scripture* (London: SCM Press, 1959), but my argument here is most influenced by the synthesis in Crouzel 1989, 91-120, 139-64; I have also referred to Karen Jo Torjesen, *Hermeneutical Procedure and Theological Method in Origen's Exegesis* (Berlin and New York: Walter de Gruyter, 1986) (hereafter cited as Torjesen) and Robert Grant, *The Letter and the Spirit* (New York: Macmillan, 1957).
7. See, e.g., Crouzel 1989, 112-20; and, of course, Henri de Lubac, *Exégèse médiévale: Les quatre sens de l'écriture,* vol. 1 (Paris: Aubier, 1959), 198-207.
8. Torjesen perhaps resolves this difficulty. She suggests that the distinction he makes between bodily and spiritual meaning corresponds to different *methods*

of interpretation, which those within the spiritual meaning do not; they correspond rather to different *effects* produced when the interpreter and his audience discover and learn the spiritual meaning (see Torjesen, 11-13 and chap. 2). Thus when he is concerned to elaborate these effects (e.g., when addressing his Christian followers in homilies or commentaries), he elaborates on one or more of the spiritual meanings; but in general, and especially when writing apology, he emphasizes the more basic difference between the methods and conclusions that belong to the bodily meaning (which he judges inadequate and potentially misleading) and those belonging to the spiritual meaning(s) (which he considers central and primary).

9. On Origen's theory of inspiration, see Crouzel 1989, 103-5. On his understanding of the Logos, and its relation to exegesis, see Crouzel 1989, 237-65, 101-6; and Torjesen, ch. 1 and 4.
10. Origen's discussion continues throughout this chapter, up to 4.3.3. He also explains that this "presence in absence" also applies to the organization of the world and the events of history: God's providence is paradoxically revealed by the ungodly and apparently improvident features of earthly life.
11. "Sensum autem divinum invenies" (e.g., *On the Song* 1.4.16); in Greek "kaì aísthesin theían eúréseis" (e.g., *Celsus* 1.38, 7.34). The Septuagint has "kai èpignosin theoú eúréseis" ("you shall find knowledge of God"; cf. the rendering in the Vulgate: "et scientiam Dei invenies"). Origen seems to have derived his idiosyncratic rendering of Prov 2:5 from Clement of Alexandria, who also judged this an important text: see *Stromata* 1.27.2 (Rahner 1932, 116-7; Harl 1975, 25-27). Harl notes that Clement finds special importance in the biblical references to *aisthésis* because he emphasizes that all knowledge comes from the senses, which she notes is an Aristotelian axiom. (I would add that Epicureans also insisted on this point long before; moreover, they seem to have developed something like a concept of spiritual senses: see above, chap. 1, n. 1). More subtly, Harl argues that Clement built on the assumptions carried by the Greek term *aisthésis* itself; Hellenistic Greeks used it to mean something close to intellectual comprehension (Harl 1975, 26-29).
12. Just before this passage Origen speculates that the mind may, like the sense organs, have an "outward shape fitly and suitably formed" for this. He thinks this hypothesis unprovable or unimaginable—apparently because he assumes mind is incorporeal—but he introduces the idea that the senses are formed by God in order to make his argument about likeness and providence. The same applies to mind.
13. Harl thinks he got the idea about sense knowledge from Clement of Alexandria, who himself seems to draw on Aristotelian teachings (Harl 1975, 26-31). On the other hand, she cautions that we often cannot be certain what Greek term the Latin *sensus* translates (Harl 1975, 30, n. 18 and 27, n. 13). On the more general principle of knowing by likeness, see, e.g., *Principles* 1.1.5, 4.4.10; touched on by Harl 1975, 23, n. 22, and Crouzel 1961, 61-63, 508-9.

14. Here again Origen's Platonic assumptions are evident. Plato teaches that we become like God by using *nous* (*mens, intellectus*) to understand *noeta* (*intelligibilia*); see *Phaedo* 66A.
15. One of Crouzel's prime conclusions (Crouzel 1961, esp. 508-13); see also David Balas, "The Idea of Participation in the Structure of Origen's Thought," in *Origeniana,* 257-75.
16. Origen assumes the already old judgment that among the senses sight is the most discerning and the most akin to intellection and, therefore, is the mechanism that can bring us away from or beyond material things to know *intellectualia,* i.e., spiritual things: see esp. Plato, *Phaedrus* 250Dff.; also *Timaeus* 46-7.
17. Cf. *Celsus* 6.69: ". . . because God is incorporeal He is not visible. But He may be perceived by those who can perceive with the heart, that is, the mind, though not with an ordinary heart, but with a pure heart [Mt 7:8]."
18. Cf. *Celsus* 7.35. A Greek version of the first sentence I quote here (Prol. 2.9), is preserved in *Catenae Graecorum Patrum in Novum Testamentum* 8 (Oxford, 1844) and reads somewhat differently: "Now just as there are these synonyms and analogous expressions, applicable both to the inner man and to the outer man, so you may find the names of the limbs of the body used metaphorically with reference to the soul" (Lawson, 27, note; the Greek can be found in Sources Chrétiennes 375, 98).
19. His immediate concern in the passage from *Heraclides* is to refute the idea that "the soul is in the blood," because this implies that the soul is material and not immortal (*Heraclides* 10).
20. Cf. *On the Song*, 3.9.4: "Haec autem propter homines qui aliter audire non possunt nisi his verbis quae in usu hebentur a Scriptura divina humani more referuntur eloquii, ut verbis quidem notis ea et solitus audiamus, sensu tamen illo quo dignum est de divinus rebus et incorporeis sentiamus." By "we" Origen clearly means not spiritual "children," who should not be reading the Song at all, or his commentary on it, but the "experienced" and "trained" whose "inner persons" are the Song's audience and referent.
21. "La formule de *Prov.* 2, 5 fournit à Origène, au singulier, le nom de la faculté globale de *perception* du divin, don't il décrit le plus souvent l'activité à travers les opérations particulières des cinq sens spirituales—et cela a cause des nombreux anthropomorphismes bibliques —, mais qu'il connait aussi pour elle-même, comme l'activité supérieure de l'intelligence humaine, du coeur de l'homme, capable de recevoir, de percevoir sensiblement, ce qui lui est directement destiné: les manifestations du Dieu qui lui est présent" (Harl 1975, 25); see also 32-34. I do not quite agree with Harl's statement that the *sensus divinus* is "l'activité supérieure de l'intelligence humaine." I think Origen makes no clear distinction between *sensus divinus* and intellect or between their operations. Harl herself admits that Origen slides imperceptibly from the activity of intellectual comprehension to the activity of sensing (Harl 1975, 32-33, n. 21) and closely relates *noeín* and *aisthésis* (Harl 1975, 26, n. 12; 33, n. 22).

22. Origen (according to Rufinus) renders Ps. 118:131 "os meum aperui et attraxi spiritum," and 2 Cor 6:11 "os nostrum patet ad vos, o Corinthii, cor nostrum dilatat est"; see *On the Song* 3.5.21; 1.1.13 and Prol. 3.4-16.
23. Here he cites verses that refer to sight, touch and, especially, eating. He typically offers a more complete list: cf. a few paragraphs after, 1.4.25, and *Celsus* 1.48.
24. Cf. *On the Song* 2.9:11-12 and *Principles* 1.1.7, on the natures of the of corporeal senses. This point is discussed by Rahner 1932, 124-25; 125, n. 85; 119-20.
25. Cf. *Celsus* 1.48; *Principles* 1.1.9. Harl emphasizes that the *sensus divinus* Origen finds in Prov 2:5 is *not* the spiritual sense of scripture, but a sense within the human person (Harl 1975, 18-25). However, she also notes that in places Origen seems to play with the term deliberately and confuse the two. Then again, it may be an artifact of translation; she cautions that *sensus* translates more than one Greek word; and Rufinus and Jerome may be suspected of using *sensus* in a way that does not exactly conform to Origen's ideas (Harl 1975, 32, n. 20).
26. This point seems confirmed by *Principles* 4.4.9 (Butterworth, 326). According to Jerome, Origen claimed that "all rational creatures . . . are of one substance. For he says, God and his only-begotten Son and the Holy Spirit are conscious of an intellectual and rational nature; so are the angels and authorities and the other powers; so, too is the 'inner man', who was made in the image and likeness of God" (*Ep. ad Avitum* 14, quoted and trans. in Butterworth, 326, n. 1).
27. He makes the same point with respect to the sense of smell: *On the Song* 1.4.21-24. On this point see Harl 1975, 22-23.
28. For my account of Origen's concept of resurrection I am especially in debt to D. G. Bostock, "Quality and Corporeity in Origen," in *Origeniana Secunda*, ed. H. Crouzel and A. Quacquarelli (Edizioni dell Ataneo, 1980), 323-37 (hereafter Bostock); Monique Alexandre, "Le statut des questions concernant la matiére dans le *Peri Archon*," in *Origeniana*, 63-81; Lawrence Hennessy, "A Philosophical Issue in Origen's Eschatology: The Three Senses of Incorporeality," in *Origeniana Quinta*, ed. Robert J. Daly (Leuven and Paris: Peeters, 1992), 373-80; Henry Chadwick, "Origen, Celsus, and the Resurrection of the Body," *Harvard Theological Review* 41 (1948):83-102 and Henri Crouzel, "Les Critiques addressés par Methode et ses contemporains à la doctrine origenienne du corps résuscité," *Gregorianum* 53 (1972):649-716; idem, "La doctrine origenienne du corps résuscité," *Bulletin de littérature ecclesiastique* 31 (1980):175-200 and 241-66 (hereafter Crouzel 1980); he summarizes his argument in Crouzel 1989, 319-31. Two works in English that extend the analysis pioneered by Chadwick and Crouzel are Jon F. Dechow, *Dogma and Mysticism in Early Christianity: Epiphanius of Cyprus and the Legacy of Origen* (Macon, Georgia: Mercer University Press, 1988), and Bynum 1995, esp. 63-68.
29. He assumes that matter is formless chaos (a belief shared by Neoplatonists, "gnostics," and Stoics; see Bostock, 324-25; Crouzel 1980, 241-46). It is distinct from both body and incorporeal intelligibles. Although it does not change

as body does (that is, into something else), it is variable and so requires a form (*eidos*).

30. My account of Origen's christology follows that of Marguerite Harl, *Origène et la fonction révélatrice du Verbe incarné* (Paris: Éditions du Seuil, 1958) (hereafter Harl 1958) and Crouzel 1989, esp. 237-55.
31. Torjesen argues that Origen "believes contemporary interpretation of Scripture to be a reenactment of the . . . teaching activity of Christ for the hearer. . . ." (Torjesen, 13). Thus the exegete who interprets the Bible correctly recapitulates Christ's teaching for himself and his hearers, in some sense taking Christ's place and himself manifesting the presence of the Logos.
32. This seems to imply that matter is uncreated and distinct from God, but Origen denies this: see e.g., *Principles* 1.3.3; 2.1.4; 2.4.3. He says further that matter will eventually be transformed and return to God with everything else: see *Principles* 3.6.8.
33. See *Hom. Lev.* 1.1. For my line of argument here, about how the spiritual person needs neither the somatic text nor the incarnate Logos, I am grateful to my colleague Amy Thomas. I paraphrase her unpublished paper, "But What About the Oxen? Body, Soul and Spirit in Origen's Exegesis," 24-25; the sentence by Origen is quoted by Thomas on p. 24, from *Commentary on John* 1.9, as rendered by A. E. Brooke, ed. and trans., *The Commentary of Origen on St. John's Gospel,* 2 vols. (Cambridge: Cambridge University Press, 1896).
34. Origen's concept of the spiritual senses had a direct and rich legacy among Greek-speaking Christians; most prominently, Gregory of Nyssa, Macarius, Diadokos of Photiké, Maximus the Confessor, Symeon the New Theologian. Little or nothing of this was known to Europeans. This history is outlined in Canévet and Fraigneau-Julien, both of which include useful bibliographies.
35. This point is made by Rahner 1932, 145, n. 239. The fragment in the *Glossa ordinaria* (at Lev. 7:5) is from Origen's discussion of the spiritual senses in *Homilies on Leviticus* 3.3. Although Origen's writings were poorly preserved in the West, significant selections, including some that discuss the spiritual senses, were preserved in monastic libraries. These were not widely cited because Origen's teachings were suspect, but monks did use them circumspectly. I shall touch on this diffuse influence in the next chapter.
36. He knew at least something of *On First Principles;* see Augustine, *City of God,* trans. Marcus Dods, George Wilson, and J. J. Smith, Modern Library (New York: Random House, 1950), 11.23; hereafter cited as *City of God.* Augustine's only substantive discussions of the spiritual senses are *Confessions,* 2 vols., ed. W. H. D. Rouse, with the translation of W. Watts (London: 1631), Loeb Classical Library (London: William Heinemann and Cambridge, MA: Harvard University Press, 1912; reprint, 1960), 10.6 and 10.27 (hereafter cited as *Confessions*); *City of God* 11.27; and *Sermon* 159.3-4, in *Patrologiae cursus completus series latina,* ed. J. P. Migne (Paris: 1844-64), 38:867-72 (hereafter cited as *PL*); also, in passing, *Tract. in Joan.* 18.5 and *Soliloquium* 1.6.12-13. Rahner notes that these three passages were often cited by later theologians, though mainly to support the idea that the spiritual senses exist,

rather than to give any details about their operation (Rahner 1932, 144-45, n. 238).

37. In some other passages where he uses sensory language to speak of how God attracts him: *Confessions* 8.4; also, more fragmentarily, 1.5, 5.1, 7.8.
38. David Chidester, "Symbolism and the Senses in Saint Augustine," *Religion* 14 (1984):31-51, and more fully in Chidester 1992, esp. 52-110.
39. See P. Landsberg, "Les sens spirituels chez S. Augustin," *Dieu vivant* 11 (1948):81-105, who unfortunately takes no account of Augustine's explicit statements about the spiritual senses and is mainly interested in Augustine's metaphysical conceptions of light, sight, and hearing.
40. This is the argument made by Carole Straw in *Gregory the Great: Perfection in Imperfection* (Berkeley and Los Angeles: University of California Press, 1988); on the spiritual senses, 39; 129. Gregory discusses the subject repeatedly: see esp. *Moralia in Job* 6.33.
41. This is demonstrated by Gregorio Penco, who lists references to the senses from throughout Gregory's writings: Penco 172-3, nn. 83-93.
42. This latter claim is my own, not Rahner's; however, my general argument about these scholastics follows his in the second part of his study of the spiritual senses Rahner 1933, 266-8.
43. *Summa aurea* lib. 4, tract. 18, q. 3, art. 2; in Guillelmus Altissiodorensis, *Summa aurea* [*Summa super quatuor libros sententiarum*], 4 vols., ed. Jean Ribaillier (Paris: Éditions du Centre national de la recherche scientifique and Rome: Editiones Collegii S. Bonaventurae ad Claras Aquas, 1980-1987), 3:506-7; hereafter cited as *Summa aurea.*
44. *De virtutibus* cap. 11; in Guilielmus Parisiensis, *Opera omnia [apud A. Pralard] 1674,* 2 vol. facsimile (Frankfurt am Main: Minerva, 1963), 1; hereafter cited as *De vitutibus.*
45. *De universibus* II, pars 2, cap. 13 in Guilielmus Parisiensis, *Opera omnia [apud A. Pralard] 1674,* 2 vol. facsimile (Frankfurt am Main: Minerva, 1963), 1:856.
46. Etienne Gilson, *History of Christian Philosophy in the Middle Ages* (London: Sheed and Ward, 1980), 258; hereafter cited as Gilson 1980.
47. Alexander discusses the spiritual senses in *Summa theologicae* pars II, inq. 4, tract. 1, sect. 2, quaes. 3, tit. 1, memb. 3, in Bernardin Klumper, ed. (Quaracchi: 1928), 2:459-64 (= no. 381); hereafter cited as *Summa* by *membrum* and chapter only: all references are to pars II, inq. 4, tract. 1, sect. 2, quaes. 3, tit. 1. We have to bear in mind that this work is a compilation, and only some of it is really by Alexander, and so it is probably safer and more accurate to attribute this treatment of the spiritual senses more loosely to the "Franciscan school" at the University of Paris in the middle third of the century, say circa 1230-1250; on the attribution problem, see Gilson 1980, 327-28 and n. 1.
48. *In III Sententiae,* dist. 13, A, art.4; in Albertus Magnus, *Opera omnia,* ed. Borgnet (Paris: Vivés, 1890-1895), vol. 28, pp. 239-41 (hereafter cited as *In III Sent.*); *De caelesti hierarchia* 15.5 (Borgnet 14:414-15) (hereafter cited as *De Cael. Hier.*). These are his two main discussions of the spiritual senses.

49. Albert's very similar discussion of the spiritual senses in *De Cael. Hier.* nicely demonstrates how he, like the other scholastics I have mentioned, prefers to cast the exegetical problem that exercised Origen as a problem in Aristotelian faculty psychology and epistemology. In *Celestial Hierachies,* Dionysius addresses the exegetical problem of sensory language. Like Origen, he asks why the sacred theologians who wrote the Bible describe angels and God himself "with forms drawn from the realm of the human" (*Celestial Hierarchies* 15.1; in *Pseudo-Dionysius: The Complete Works,* trans. Colm Luibheid [New York: Paulist Press, 1987], 184). He answers that these images, and also the actual "variegated forms and shapes adopted by angels," allow us to "take off from these images, and . . . by retracing, rise up again to the simplicity of heavenly minds" (*Celestial Hierarchies,* 182). In other words Dionysius, in accordance with his hierarchical cosmology, interprets anthropomorphisms, including sensory language, as created signs. Albert almost completely ignores Dionysius' exegetical point. He expands on Dionysius' discussion of how angels "sense," then glosses these anthropomorphisms as aspects of the human person—specifically, as acts of the intellect. For "[quid] dicitur sensus in spirituali natura [i.e., in angelis], intellectus rei praesentis: propter similem modum accipiendi ad sensum" (*De Cael. Hier.* 15.5).
50. Albert's two accounts agree almost exactly, except that in *De cael. hier.* 15.5, dubium he also gives Dionysius' alternative ranking of the senses, which after sight puts not hearing but smell. This is puzzling, because elsewhere Albert echoes Dionysius, writing that the highest name of God is ***bonum,*** not ***verum.*** Perhaps the confusion has to do with Albert's idea that smell in some sense perceives both: "verum in ratione boni" (in *III Sent.,* as I discuss in the previous paragraph).
51. I pass over Albert's student Thomas Aquinas (1224-1274) who touches on the idea of the spiritual senses but, as Rahner observes, adds nothing significant to Albert's ideas (Rahner 1933). The key passages are as follows: *III Sent.* dist. 13, expos.; *In Philipp.* cap. 2.1.2; *In Psalm* 33, n. 9. On Aquinas, see Rahner 1933, 268 and nn. 28 and 29; in connection with the senses of angels, Rahner also mentions Anthony of Padua (1195-1231).

Notes to Chapter 3

1. Bernard seems to have known the Latin text of Origen's *Commentary on the Song of Songs,* which discusses and applies the concept; there are a number of very close parallels. See Jean Leclercq, "Aux sources des sermons sur les Cantiques," *Receuil d'études sur saint Bernard et ses écrits* (Rome: Edizioni di storia et letteratura, 1962-1969), 1:275-319, esp. 281-83 (hereafter cited as Leclercq 1962); Jean Prosper Theodorus Deroy, *Bernardus en Origenes: Enkele opmerkingen over de invloed van Origenes op Sint Bernardus* (Haarlem: De Toorts, 1963); and Gustave Bardy, "Saint Bernard et Origène," *RMAL* 1 (1945):420-21. Jean Leclercq has shown that in the twelfth century, the *corpus Origenianum* was held by and diffused from the Cistercian libraries

of Clairvaux, Signy, Pontigny, and St. Thierry; see "Origène au XIIe siècle," *Irenikon* 24 (1951):425-39. For a brief summary of Leclercq's own conclusions and others', see Jean Leclercq, *The Love of Learning and the Desire for God: A Study of Monastic Culture,* 3rd ed., trans. Catherine Misrahi (New York: Fordham University Press, 1982), 94-97 (hereafter cited as Leclercq 1982); more recently, Michael Casey has summarized scholarship on Bernard's written sources in *Athirst for God: Spiritual Desire in Bernard of Clairvaux's Sermons on the Song of Songs,* Cistercian Studies 77 (Kalamazoo, MI: Cistercian Publications, 1988), 22-32 (hereafter cited as Casey). Further evidence that the doctrine was known among Cistercians is the explicit mention of spiritual senses in the anonymous Cistercian treatise *De spiritu et anima* (chapters 9 and 49), which was compiled perhaps twenty years after Bernard's death (c. 1170).

2. Several scholars have noted the tendency and discussed aspects of Bernard's use of sensory language. Among the major scholars, see Gilson 1990, 91; Jean Mouroux, "Sur les critères de l'expérience spirituelle d'apres les Sermons sur le Cantique de Cantiques," in *Saint Bernard Théologien* (Analecta Cisterciensia 9), ed. Jean Leclercq (Rome: Editiones Cistercienses, 1953), 250, 264, and esp. 258-60 on hearing, touch and sight (hereafter cited as Leclercq 1953); Jean Leclercq, "De quelques procédés du style biblique du S. Bernard," in Leclercq 1962, 3:263-65; Leclercq 1982, 30-32, 60-61; idem, *Bernard of Clairvaux and the Cistercian Spirit,* trans. Claire Lavoie (Kalamazoo, MI: Cistercian Publications, 1976), 22-23, 151-52 (hereafter cited as Leclercq 1976); and Casey, esp. 296-98, 231-34 (on sight and hearing), and 79-80 (on taste and eating). Among other scholars, see Bernard Bonowitz, "The Role of Experience in the Spiritual Life," in *La dottrina della vita spirituale nelle opera di San Bernardo di Clairvaux: Atti del Convegno Internazionale, Roma, 11-15 settembre 1990,* Analecta Cisterciensia, 46 (Rome: Editiones Cistercienses, 1990), 323. The spiritual senses are at least hinted at in the studies of Edith Scholl, "The Sweetness of the Lord: Dulcis and Suavis," *Cistercian Studies Quarterly* 27 (1992):359-66, and Franz Posset, "Christi Dulcedo: The 'Sweetness of Christ' in Western Christian Spirituality, *Cistercian Studies Quarterly* 30 (1995):245-65. Susan Warrener Smith discusses Bernard's sensory metaphors in her article "Bernard of Clairvaux and the Nature of the Human Being: The Special Senses," *Cistercian Studies Quarterly* 30 (1995):3-13, but does not directly address the concept of the spiritual senses, because she is primarily concerned with exploring the history of natural theologies that affirm that the physical world is a good creation.

3. Bernard McGinn has shown that Bernard's sensory language is well integrated with his mystical theology and noted that it implies a concept of the spiritual senses. However, McGinn assumes that Bernard's concept is a version of that we find in Origen, Augustine, and Gregory the Great (and so assumes that we can trace a continuous and coherent "doctrine" of the spiritual senses in Christian mystical texts). See *The Growth of Mysticism: Gregory the Great*

through the Twelfth Century (New York: Crossroad, 1994), 185-90; hereafter cited as McGinn 1994.

4. There are many studies on ideas of the individual and/or the self in this period. Two especially influential and clearly argued examples are Colin Morris, *The Discovery of the Individual, 1050-1200,* Medieval Academy of America Reprints for Teaching 19 (Toronto: University of Toronto Press, 1987; first published 1972) and Caroline Walker Bynum, "Did the Twelfth Century Discover the Individual?" in *Jesus as Mother: Studies in the Spirituality of the High Middle Ages* (Berkeley and Los Angeles: University of California Press, 1982) 82-109 (Bynum's later books are also relevant; her published works can be read as an extended exploration of medieval conceptions of self).
5. *Sermones super Cantica Canticorum* 6, in *Sancti Bernardi Opera Omnia,* 8 vols., ed. J. Leclercq, C. H. Talbot, and H. M. Rochais (Rome: 1957-80), 1:27 (hereafter cited as *SBO*). I cite Bernard's works by section, and give a reference to the volume and page in *SBO* for quotations. For works that I cite only once or twice, I give the full title; for works I cite frequently, I use the standard abbreviations, as follows: *Conv., Ad clericos de conversione; Csi., De consideratione libri V; Dil., De diligendo Deo; Div., Sermones de diversis; Ep., Epistolae; QH, Sermo super Psalmum "Qui habitat"; SC, Sermones super Cantica Canticorum; Sent., Sententiae.* Most translations are my own, but I have benefitted from those in the Cistercian Fathers series, which I have used for some passages and cite at first use.
6. Bernard is generally viewed as one of the great champions of the newly fervent devotion to the physical Christ. Richard Southern has well observed that he did not invent this type of piety: "Long before theory caught up with practice the sufferings of Christ had excited the pity of unlettered men, who knew nothing of the theology of Redemption which made pity irrelevant" (*The Making of the Middle Ages* [New Haven: Yale University Press, 1959], 256). It is useful to think of Bernard as an integrator: his writings and his influence as the leading light of the Cistercian order helped cement personal devotion to the man Jesus (and his mother) firmly into the thought and practice of medieval Christianity, both popular and learned.
7. On Bernard's doctrine of bodily resurrection, and its relation to the ascension of Christ, see Casey, 234-41; Bernard McGinn, "Resurrection and Ascension in the Christology of the Early Cistercians," in *Citeaux in de Nederlanden,* 30 (1979):5-22, and Bynum 1995, 163-66.
8. More broadly, Jean Leclercq has emphasized that for Bernard, as for other monks, the Bible was "an experience"—that is, *practice* (Leclercq 1976, 22-23).
9. Bernard expands on experience (among other places) in *SC* 1.9-10 and *SC* 3-4; see also *SC* 56.1. On Bernard's concept of experience, see B. Mouroux, "Sur les critères de l'expérience spirituelle d'apres les Sermons sur le Cantique de Cantiques," in Leclercq 1953, 251-67; Brian Stock "Experience, Praxis, Work and Planning in Bernard of Clairvaux: Observations on the *Sermones in Cantica,*" in *The Cultural Context of Medieval Learning,* ed. with an intro-

duction by J. E. Murdoch and E. D. Sylla, (Dordrecht: Reidel, 1976), 219-68 (hereafter cited as Stock 1976); Ulrich Köpf, *Religiöse Erfahrung in der Theologie Bernhards von Clairvaux* (Tübingen: Mohr, 1980), summarized in "Die Roller der Erfahrung im religiösen Leben nach dem heiligen Bernhard von Clairvaux," in *La dottrina della vita spirituale nelle opera di San Bernardo di Clairvaux,* Analecta Cisterciensia 46 (Rome: Editiones Cistercienses, 1990); and, in the same volume, the reply to Köpf by Bernard Bonowitz, "The Role of Experience in the Spiritual Life," 321-25.

10. *De gradibus humilitatis et superbiae* 6-12 (*SBO* 3:20-26).
11. Translation is from *On the Song of Songs,* 4 vols., trans. Kilian Walsh and Irene Edmonds (Kalamazoo, MI: Cistercian, 1971-80) 3:88; hereafter cited as OSS.
12. E. von Ivánka, "L'union à Dieu: La structure de l'âme selon S. Bernard," in Leclercq 1953, 204. To express this idea, Bernard adopts at least three two-fold and three-fold "patterns" for "states" of the soul. These patterns are not consistent. According to Von Ivánka, the only distinction Bernard always maintains is that between "the two arms of the soul": understanding and will, or knowledge and love (von Ivánka, 206).
13. A good deal of attention has been devoted to Bernard's biblical rhetoric. Etienne Gilson discusses the subject in *The Mystical Theology of Saint Bernard,* trans A. H. C. Downes (Kalamazoo: Cistercian Publications, 1990), esp. 99-118, 144-46. Leclercq wrote a number of important studies on Bernard's language and style, many collected in Leclercq 1962, esp. vol. 3. He stresses that Bernard's Latin was "Christian . . . and not secular" (Leclercq 1976, 31-32); his sources and his language are "biblical," "liturgical," and "Patristic" (Leclercq 1976, 22-25, 29-34). Christine Mohrmann further elaborates this point with evidence from *the Sermons on the Song* in her essay "Observations sur le langue et le style de saint Bernard," in *SBO* 2:ix-xxxiii. Leclercq also makes the more precise claim that Bernard used key words in their biblical senses, quite aware of their precise (etymological?) meanings: see "La Bible dans les hómelies de S. Bernard sur "*Missus est,*" in Leclercq 1962, 2:213-48, esp. 229-37; and on "[-]*spirare,*" Leclercq 1976, 33; on "*comprehendere*" as it occurs in *Csi.* 5.XIII.27-30, ibid., 151-52). See also Jacques Blanpain, "Langage mystique, expression du désir dans les Sermons sur le Cantique des cantiques de Bernard de Clairvaux," *Collectanea Cisterciensia* 36 (1974), 46. See also M. Dumontier, *S. Bernard et la Bible* (Bruges: 1953). The classic treatment of monastic practice and theology is Leclercq 1982.
14. Song 1:3, *SC* 21 and 22; Song 2:9-10, *SC* 57. We might ask whether the class "sensory metaphors" can be strictly segregated from the erotic themes and motifs that permeate Bernard's writings. This question seems particularly pertinent because Bernard draws both sorts of language from the erotic drama of the Song. And of course our answer must be that no such neat distinction is possible. Yet I would argue that we should treat any example as sensory language first, erotic second, in order not to overlook how the language of the text works *such that we interpret it* as erotic.

15. Denys Turner has recently made a provocative argument about why medieval exegetes took up the Song with such enthusiasm: see *Eros and Allegory: Medieval Exegesis of the Song of Songs* (Kalamazoo, MI: Cistercian Publications, 1995); hereafter cited as Turner.
16. This is a major strand in Turner's argument: see esp. 85-89, 127-34 and 154-5.
17. *SBO* 6,1:122-24; 393-94; 6,2:108-112. The authenticity and interpretation of these sermons present some problems: see *SBO* 6,1, "Introduction," 59-71. Another passage where Bernard may also have something like the spiritual senses in mind is Sent 3:70 (*SBO* 6,2: 102-7), where he discusses the "spirit of wisdom and understanding" that rests in Christ (Is 11:1-3) by likening the mediator Christ to the "mediating" sense, taste. This is a suggestive passage but does not offer an explicit doctrine of spiritual senses.
18. Note that, although he does not draw it out, Bernard implies that sight and hearing are more perfect in humans than in animals, because he assimilates them to intellection and understanding. Bernard echoes his treatment in this sermon in *SC* 50.8 (*SBO* 2:83-4; OSS 3:36-37).
19. However, his friend William of St. Thierry did develop the idea that love is the sense of the soul and gave it a strong trinitarian interpretation. In *De natura et dignitate amoris* 6, 7, 10 (in *PL* 184:379-408), and again in *Speculum fidei* 15 (in *PL* 180:365-87), William reproduces the treatment we find in *Div.* 10, and perhaps we can assume that he picked up the idea from Bernard. See James Walsh, "William of St.-Thierry and the Spiritual Senses," *Revue d'ascetique et de mystique* 35 (1959):27-42; David N. Bell, *Image and Likeness: The Augustinian Spirituality of William of St. Thierry* (Kalamazoo, MI: Cistercian Publications, 1984), esp. 161-65 and ch. 6 (hereafter cited as Bell); M. M. Davy, *Théologie et Mystique de Guillaume de St.-Thierry: I. La connaissance de Dieu* (Paris: Vrin, 1954), esp. 222-52.
20. The sentence in question reads as follows: ". . . homo interior . . . in sensu autem discernitur in quinque partes notissimas, scilicet, visu, gustu, auditu, odoratu et tactu, quia aliter sentis in oculo, aliter in *aure,* et sic in ceteris. . . ." (*SBO* 6,1:394; emphasis mine). One manuscript has "ore" for "aure," which would correspond better with the list in the earlier part of the sentence. Either the order was not important to Bernard, which I consider unlikely, or we have only corrupt manuscripts. The most logical assumption is perhaps that an early scribe transposed "gustu" from fourth position to second. As we have it now, the order of the senses also does not correspond well with the lists of *invisibilia Dei* and qualities *propter quam diligitur*. This is because elsewhere Bernard consistently teaches that truth—at least the truth we can attain in this life—corresponds not to sight but to hearing. Likewise he usually links wisdom, sweetness, and taste, as I shall discuss more below.
21. This is precisely the distinction William of St. Thierry makes in *De natura et dignitate amoris* 6 and 7 (*PL* 184:391A-393A).
22. This was the standard theory of sight in the Latin West until the late twelfth century. Although derived ultimately from ancient Greek philosophers, in the

Latin West these sources were scarcely known, except the discussion of sight in Plato's *Timaeus;* the theory was mainly known from Augustine's discussion in Book 12 of *De Genesi ad Litteram* and Book 11 of *De Trinitate.* On Augustine's analysis of vision and visions, see Chidester and McGinn 1991, chapter 7, esp. 253-56. Karl Morrison makes some perceptive remarks about Bernard's "equation between seeing and loving" in his article "Hermeneutics and Enigma: Bernard of Clairvaux's *De consideratione,*" *Viator* 19, esp. 143-44.

23. Cf. the immediacy of *unitas spiritus* that Bernard discusses in *Sermon 2* on the Song, a "kiss" (which he glosses as "adhering of the lips") that surpasses ". . . visions and dreams, . . . parable and figures of speech . . . [and] the beauty of the angels . . ." (*SC* 2.2; cf. *SC* 5.8-9, *SC* 8).
24. McGinn 1994, 187-88. It is not clear whether McGinn means subjectively or objectively (rhetorically) synaesthetic.
25. See also *SC* 37.4 (*SBO* 2:11), where Bernard writes that he who has been "borne aloft on the wings of grace" has had a vision of God, heard his voice, and tasted and seen that God is sweet. Likewise he writes that "a vision as clear as it is everlasting will steal upon her, bringing not only sweetness to her taste but fulfillment to her heart. . . ." (*SC* 48.8; *SBO* 2:72-73); and that monks seek "insight that they may grasp what they long for, wisdom in order to savor what the mind grasps" (*SC* 9.3; *SBO* 1:44). Similarly, *SC* 85 is structured around "the wisdom that is tasted" but also "the light of truth that shines in the mind."
26. E.g., *SC* 31.1-2 (*SBO* 1:219-20), where he makes clear that heavenly sight is immediate. Not so earthly vision. While *visio Dei* in the broad sense was considered possible in this life, medieval theologians always qualified and limited it. They relied heavily on Augustine's analysis of vision in *De genesi ad litteram* 12, he teaches that there were three forms of *visio Dei* in this life: corporeal, spiritual or imaginary, and intellectual. All are brief and all mediated.
27. Bernard's frequent term for this sampling is *prelibatio,* a reference to taste. Another is *contemplatio,* which is a complex term but basically a reference to sight; he often characterizes it with language of touch or taste.
28. E.g., *SC* 41; *SC* 28; *SC* 56; *QH* 8 (*SBO* 4:432; CF 25:169-81). Bernard's debt to Augustine is perhaps clearest in *On Conversion.* For some remarks on this pattern, see Duncan Robertson, "The Experience of Reading: Bernard of Clairvaux *Sermons on the Song of Songs,* I," in *Religion and Literature* 19 (1987), esp 9-13; on Augustine see Chidester.
29. Although Bernard bases the primacy of spiritual hearing on biblical verses, Chidester has shown that a similar relation between hearing and sight is a major motif in Augustine's writing, which no doubt influenced Bernard.
30. Although Bernard does also call grace a vision he longs to see, quoting Ps 41:3, "When shall I see the face of God?," here again he seems to be referring to the beatific vision, the perfected heavenly *visio Dei.*

31. The translations of this and the next passage are from *Five Books on Consideration: Advice to Pope,* trans. John D. Anderson and Elizabeth T. Kennan (Kalamazoo, MI: Cistercian, 1976), 177, slightly modified.
32. Sometimes *tangere,* more often *amplexus, osculum,* and *adhaesio*: e.g., *Sermones in Quadragesima* 6.3 (*SBO* 4:365, l.16): "Quae vero mundus reputat crucem, illia affixus sum, illis adhaero, illa toto amplector affectu."
33. My account here is based on the detailed discussion in Casey, 94-116; Casey gives many citations. Cf. Bell, 125-35, who discusses how Bernard's friend William of St. Thierry uses the term.
34. Bernard often identifies them: *affectus, id est voluntas* (*Sermo in ascensione Domini* 3.2), as noted in Casey, 101).
35. E.g., *Sermo in feria iv hebdomadae sanctae* 14. See also *SC* 3.5 and *SC* 83.3 and 83.6.
36. See *Sermones in nativitate Domini* 2.6 (*SBO* 4:256); also *Sermones in adnuntiatione dominica* 3.8 (*SBO* 5:40), *Div.* 92.1 (*SBO* 6,1:347), and *SC* 71.10, where Bernard describes the adhering of the person to God as a mutual *in*hering, one in the other.
37. Variants include "gustat et sapit," *SC* 50.4 (*SBO* 2:80; "gustare et sentire," *SC* 19.7 (*SBO* 1:112); "gustavimus et scimus," *Dil.* 9.26 (*SBO* 3:141).
38. *In nativitate beatae Mariae Virginis* 2.
39. See also *Dil.* 15.39; *Ep.* 11:8; also *Homiliae super 'Missus est'* 3:6 (*SBO* 4:39), where Bernard says that those who taste and are transformed stimulate others by "belching" forth praise.
40. In addition to the passages I touch on here, other substantial passages built around the words *sapere, sapor,* and *sapientia* are *SC* 67.6, *SC* 23.14ff, and *Sent.* 3:96. The significance of this group of words is briefly discussed by Casey, 297-8. Bernard's taste for and skill at this kind of word-play is analyzed by Leclercq in "Sur le caractère litéraire des sermons de S. Bernard," in Leclercq 1962, 3:163-210.
41. Translation from *Sermons on Conversion: On Conversion, a Sermon to Clerics, and Lenten Sermons on the Pslam 'He Who Dwells,'* trans. with an introduction by Marie-Bernard Saïd, OSB (Kalamazoo, MI: Cistercian Publications, 1981), 60-61.
42. The metaphor of instruction as eating that is tasted seems as old as language, but Deroy, in *Bernardus en Origenes,* has shown that Bernard's treatment of this theme in *SC* 23 closely follows Origen's. See also Leclercq 1962, 3:75-6.
43. If we think about it phenomenologically, following Jonas, it seems clear that the act of eating goes beyond immediate contact, to engulfing and assimilating. Therefore references to eating seem to me conceptually distinct from references to tasting. In these paragraphs, I think Bernard recognizes this. I will come back to the subject in the next chapter with reference to Hadewijch, and make a few remarks about scholarly approaches to language of eating.
44. Here Bernard may be echoing Augustine, *Confessions* 7.10, who mentions eating without consuming.

Notes to Chapter 4

1. Hadewijch's dates present something of a problem. We know nothing about her aside from what she tells us in her own writings, and she tells us next to nothing. Scholars now generally agree she wrote in the mid-thirteenth century, probably in the years 1230-1250. This date is based largely on philological arguments and on her references to some people known from other records. The most recent summary of scholarship on Hadewijch's dates and on the evidence of the surviving manuscripts is given in Reynaert 1981; for a brief summary in English, see Saskia Murk-Jansen, *The Measure of Mystic Thought: A Study of Hadewijch's Mengeldichten* (Göppingen: Kümmerle Verlag, 1991), 9-14, 21-24, and idem, "The Mystic Theology of the Thirteenth-Century Mystic, Hadewijch, and Its Literary Expression," in *The Medieval Mystical Tradition in England 5*, ed. Marion Glasscoe (Woodbridge, Sussex and Rochester, NY: D.S. Brewer, 1992), 117-27, esp. 118-20 (hereafter cited as Murk-Jansen 1992). Recently Wybren Scheepsma has argued, not entirely convincingly, that Hadewijch was active around 1300, and identifies her with the "Heilwig Bloemardinne" that Ruusbroec attacks in his writings: "Hadewijch und die *Limburgse Sermonen* Überlegungen zu Datierung, Identität und Authentizität," "Hadewijch und die *Limburgse Sermonen:* Überlegungen zu Datierung, Identität und Authentizität," in *Deutsche Mystik im abendländischen Zusammenhang: Neu erschlossene Texte, neue methodische Ansätze, neue theoretische Konzepte—Kolloquium Kloster Fischingen 1998,* edited by Walter Haug, 653-82 (Tübingen, 2000). (Thanks to Bernard McGinn for drawing this paper to my attention).
2. This point is emphasized by Bynum 1988, 153-60; and by Milhaven, esp. 101-110; see also Jantzen, 134-46. Partly because of the popularity of Bynum's argument, "eucharistic piety" has become a boilerplate characterization that people too often apply to any and all thirteenth-century holy women, without properly weighing other themes and ideas in the surviving sources. Although Hadewijch certainly writes about the Eucharist, I think it is less central to her ideas than many of Bynum's readers might conclude.
3. Leonce Reypens, who long ago examined what he called "mystic touch" as it is discussed by appears in the writings of Jan Ruusbroec, notes that he seems to have picked it up from Hadewijch, though he also finds traces of it in Beatrijs of Nazareth's *Van Sieven Manieren van Minne.* ("Ruusbroec-studien. I. Het mystieke 'gherinen'," *Ons Geestelijk Erf* 12 [1938]:15-86.) He notes a significant allusion at the end of Hugh of St. Victor's *Soliloquium de arrha animae* (Reypens, 158-59, and nn. 1-4; the passage from Hugh can be found in *PL* 176:970C). However, it seems to be the beguine theologians who developed language of touch to discuss union with God as a dynamic trinitarian bond.
4. In the surviving manuscripts and in modern editions Hadewijch's writings appear in four groups, which roughly correspond to literary genres. (1) There are twenty-nine or thirty didactic letters addressed to unnamed followers (hereafter cited as *Letter* by number and line). These are generally presented as

prose, but they also include sections of poetry in couplets, and are thick with other rhythmic, rhyming passages. (2) There is a fourteen-chapter *Visioenenboek* (*Book of Visions,* hereafter cited as *Visions* by chapter and line) that presents a unified narrative of spiritual growth. It includes a "List of the Perfect" (hereafter cited as *List* by line) that includes biblical characters, such as Mary and John the Baptist; church authorities, such as Augustine and Bernard of Clairvaux; anonymous holy women known to Hadewijch; and others. (3) There are forty-five *Strofische Gedichten* (*Poems in Stanzas,* hereafter cited as *PS* by number and line), whose form and motifs are almost indistinguishable from those we find in contemporary Northern French poetry of *fine amour.* (4) There is a collection of *Mengeldichten* (*Poems in Mixed Forms,* hereafter cited as *PM* by number and line), of which some seem to be letters and others doctrinal treatises. I cite these writings from the critical editions prepared, with extensive introductions, commentary, and notes, by Jozef van Mierlo: *Hadewijch: Visioenen,* 2 vols. (Louvain: Vlaamsche Boekanhalle, 1924, 1925); *Hadewijch: Brieven,* 2 vols. (Antwerp: Standaard, 1947); and *Hadewijch: Mengeldichten* (Antwerp: Standaard, 1952). The exception is the *Stanzaic Poems,* which I cite from a more recent edition that reproduces Van Mierlo's Middle Dutch text (for which see the Bibliography): *Hadewijch. Strofische Gedichten,* ed. and trans. (into Modern Dutch) by E. Rombauts and Norbert de Paepe (Zwolle: Tjeenk Willink, 1961); hereafter cited as de Paepe 1961. There is an English translation of the whole collection, though it is not reliably precise: *Hadewijch: The Complete Works,* trans. with an introduction and notes by Mother Columba Hart, O.S.C. (New York: Paulist Press, 1984); hereafter cited as Hart. There is a more recent translation of the *Poems in Stanzas:* Marieke van Baest, trans., *Poetry of Hadewijch* (Leuven: Peeters, 1998). I list other translations and recent Dutch editions in the bibliography. All translations in the text are my own, except as noted, but I have benefitted from Hart's translation and other English, French, and Modern Dutch translations.

5. Scholars have been writing about beguines since about the time their name was made popular in a song. For a brief, clear, and balanced recent overview and a well-chosen bibliography, see Saskia Murk-Jansen, *Brides in the Desert: The Spirituality of the Beguines* (Maryknoll, NY: Orbis Books, 1998); hereafter cited as Murk-Jansen 1998.
6. Even literate medieval people cultivated prodigious memories. Mary Carruthers has explained how in *The Book of Memory: A Study of Memory in Medieval Culture* (Cambridge: Cambridge University Press, 1990) and argued convincingly that we should recognize that the techniques, the tropes, and the idea of memory were more fundamental to medieval culture than were those of the book.
7. In these respects, the literary culture of medieval women who sought religious education and a holy life differed significantly from that of male clerics, which is one possible reason that women who wrote about God tended to write, and perhaps think, in other patterns than did men.

8. We can only guess how Hadewijch came by her education, on the basis of other accounts, such as that in Beatrijs of Nazareth's *vita*: see *The Life of Beatrice of Nazareth (1200-1268)* trans. and annotated by Roger DeGanck (Kalamazoo, MI: Cistercian Publications, 1991), xiii-xvii and 24-27. See also the discussion by Herbert Grundmann in his study *Religious Movements in the Middle Ages,* trans. Steven Rowan, (Notre Dame: University of Notre Dame Press, 1996), 192-201.
9. Joris Reynaert, "Mystische bibelinterpretation bei Hadewijch," in *Grundfragen Christlicher Mystik,* ed. M. Schmidt and D. Bauer (Stuttgart-Bad Canstatt: Frommann-Holzboog, 1987), 123-37 and his complementary discussion in Reynaert 1981, 403-35. See also Ward Corsmit, "De Heilige Schrift bei Zuster Hadewijch," *On Geestelijch Leven* 40 (1963):151-60.
10. She heavily favors a few books (Job, John, and Matthew) and refers more than once to only a few texts. Reynaert gives a handy chart of his conclusions regarding her biblical references in Reynaert 1981, 411-13, and also lists and briefly analyzes her more sustained paraphrases and commentaries, 414-23.
11. Nor, for that matter, does Hadewijch make much use of the Song of Songs itself. Reynaert identifies perhaps 25 references, but eleven of these are to a single verse, 2:16 ("delectus meus mihi et ego illi"); she refers to only six other verses: 1:1 (4 times), 1:2 (3 times), 1:3, 2:12, 3:4, 7:12 (2 times). None of these references support any sustained dissection of love or the relation between lovers; they are mere proof-texts.
12. Barbara Newman, "*La Mystique Courtoise:* Thirteenth-Century Beguines and the Art of Love," in *From Virile Woman to WomanChrist: Studies in Medieval Religion and Literature* (Philadelphia: University of Pennsylvania Press, 1995), 167-37; on this point see esp. 138-39, 143-48; hereafter cited as Newman. There are two other clear discussions in English: Murk-Jansen 1998, 34-58; and Joris Reynaert, "Hadewijch: Mystic Poetry and Courtly Love," in *Medieval Dutch Literature in its European Context,* ed. Erik Kooper (Cambridge: Cambridge University Press, 1994), 208-25; hereafter cited as Reynaert 1994.
13. There is an forest of scholarship on medieval literary conventions and their medieval and modern interpretations. Three studies that I have found helpful are Sarah Kay, *Subjectivity in Troubadour Poetry* (Cambridge and New York: Cambridge University Press, 1990); Kate Greenspan, "Autohagiography and Medieval Women's Spiritual Autobiography, in *Gender and Text in the Later Middle Ages,* ed. Jane Chance, 216-36 (Gainesville: University Press of Florida, 1996), who specifically discusses the issues of gender and genre; and, with reference to mystical writings, Stephanie Paulsell, "Scriptio Divina: Writing and the Experience of God in the Works of Marguerite d'Oingt" (Ph.D. dissertation, University of Chicago, 1993).
14. In addition to the examples of this approach that I discuss in the following paragraphs, see also Ulrike Wiethaus, "Learning as Experiencing: Hadewijch's Model of Spiritual Growth," in *Faith Seeking Understanding,* ed. G. Berthold, (Manchester, NH: St. Anselm College Press, 1991), 89-106; and Paul

Mommaers, "Functie van der taal in de mystieke beleving volgens de *Brieven* van Hadewijch," *Ons Geestelijk Erf* 61 (1987):135-62; see also the introductory chapters to Mommaers' edition of *De visionen van Hadewijch* (Nijmegen: 1979), esp. 193-204.

15. Jantzen similarly claims that Hadewijch, like other "women mystics," wrote about her own embodied experience. For example, she writes that Hadewijch does not merely use erotic imagery as "allegory" for spiritual relation to God, but "treats the erotic . . . as an actual, though visionary, encounter" (135; on Hadewijch, 134-46, 164-65; Jantzen does not flesh out what she means by "actual").
16. Milhaven does concede, reluctantly, that the real issue is not "what happened" (i.e., whether or not Christ really *did* interact physically with Hadewijch) but, rather, the fact that in places Hadewijch *presents* her knowledge as gained through physical, bodily interactions (89). Yet Milhaven obviously is not satisfied with this cautious claim about Hadewijch the writer. Although he wants to avoid making claims about what *God* did or did not do, he very much wants to make claims about what medieval women experienced, because his argument is that they embraced and championed embodied experience. Otherwise he has no evidence for the claim he wants to make: that "Hadewijch experienced *herself as physically embracing her Beloved*" (15); and that Marguerite d'Oingt ". . . experienced the experience as of a bodily kind" (103).
17. Hadewijch examines the interplay between these themes in *Letter* 6, which is her longest sustained examination of willing *minne*/God. She emphasizes that "to live sincerely according to the will of *minne* is to be so perfectly one in the will of true *minne* . . . that even if one had another wish, one would choose or wish nothing except to desire above all what *minne* wills, no matter who is condemned or blessed by it" (6.76-82; translation from Hart, 58).
18. Amy Hollywood, *The Soul as Virgin Wife: Mechthild of Magdeburg, Marguerite Porete and Meister Eckhart* (Notre Dame and London: University of Notre Dame Press, 1995), esp. ch. 2, which argues this point at length, noting that Bynum uses sources without addressing this crucial difference of authorship and intent.
19. Van Mierlo believed she wrote for her confessor, and Stephanus Axters agreed with him (*The Spirituality of the Low Countries,* trans. Donald Attwater [London: Blackfriars, 1954], 90-95). But more recent scholars have considered it unlikely: see H. Vekeman "Angelus sane nuntius: Een interpretatie van het Visioenenboek van Hadewijch," *Ons Geestelijk Erf* 50 (1976):225-59; J. Porion, trans., *Visions,* (O.E.I.L: 1986), p. 86, n. 4 (who also mentions a corroborating remark by Romana Guarnieri, but gives no citation); and the summary remarks in Murk-Jansen 1998, 35-36, 41-42.
20. On Hadewijch's use of the word *sin,* see Saskia Murk-Jansen, "The Use of Gender and Gender-Related Imagery in Hadewijch," in *Gender and Text in the Later Middle Ages,* ed. Jane Chance, (Gainesville: University Press of Florida, 1996), 61-62; hereafter cited as Murk-Jansen 1996.

21. There is a sentence in *Visions* 3 that might refer to the interior senses, but I think the mind: "Later, one Easter day, I had gone to God; and he embraced me from within my mind and took me away in the spirit" (*Visions* 3:2-3). [Daer na eens paesechs daghes wasic te gode ghegaen; ende hi omvinc mi van binnen mine sine ende nam mi inden gheest. . . .]
22. Hadewijch mentions "De heilige Bernardus" in the *List of the Perfect* (*List* 144), although she continues "over hem weet ik ook wenig." This is no proof she knew Bernard's actual writings first hand. She may well be referring to the works of William of St. Thierry, which she clearly knew, and were transmitted under Bernard's name. Nevertheless, Bernard's influence was ubiquitous. With reference to sensory language, I note some ideas and phrases common to both. Both speak of God and person eating each other (I discuss this below, with citations, p. 95; cf. my discussion with respect to Bernard, above, pp. 64-5 and Ruusbroec, below, p. 115). Both also mention the notion of *gluten amoris,* traditionally associated with I Cor 6:17 (Hadewijch: *Letter* 16:30-32, *Letter* 22:58; Bernard: e.g., *SC* 71:8, *Epist.* 266:2. I might also note the more tenuous association between touching, grip, and I Cor 6:17 (Hadewijch: *Letter* 22:120-21; Bernard: *SC.* 8.2, 71.8; 79; *Div.* 4.3).
23. Along with the word "*minne*" and its derivatives, Hadewijch also uses the word "*lief(e)*." Although these words are usually analyzed together and are usually not careful distinguished in translations, I think Hadewijch uses them somewhat differently. She uses "*minne*" as an ambiguous and multiplex term to highlight the essential bond between creation and God; she does not use *lief(e)* that way. An especially interesting piece of evidence for this is a juxtaposition in *PM* 15, where Hadewijch writes consecutive stanzas that play separately on *lief* and *minne:* the first (*PM* 15:45-48) emphasizes the actions of the person and God as discrete beings, the second (*PM* 15:49-52) the state of unified beings.
24. Until fairly recently most scholars have tried to shoehorn Hadewijch's conception of *minne* into a clearly circumscribed definition, at least partly in order to hem in its possibly unorthodox implications. Van Mierlo argued that *minne* is God or Christ ("De 'Minne' in de Strophische Gedichten," *Verslagen en Mededelingen van de Koninklijke Vlaamse Academie voor Taal- en Letterkunde,* 1941, 687-705); de Paepe that it is an entity outside God that expresses Hadewijch's experience of God (de Paepe 1961, 261); DeGanck has an even more experiential interpretation in his *Beatrice of Nazareth in Her Context* (Kalamazoo, MI: Cistercian Publications, 1991), esp. 471-73. More recently some scholars have tried to escape these too-restrictive options and stressed the essentially ambiguity of *minne*: there is a large literature on the significance of term in medieval German literature, but with reference to Hadewijch in particular, see Reynaert 1981, 333-61; Frank Willaert, *De poëtica van Hadewijch* (Utrecht: 1984), 347-57; Kurt Ruh, "Gottesliebe bei Hadewijch, Mechthild von Magdeburg und Marguerite Porete," in *Romanische Literaturebeziechungen im 19. und 20. Jahrhundert,* ed. A. San Miguel *et al.,* (Tübingen: Gunter Narr Verlag, 1985), 243-54, esp. 272-74;

idem, *Geschichte der Abendländischer Mystik*, vol. 2, *Frauenmystik und Franziskanische Mystik der Frühzeit* (Vienna: Mohr, 1993), 155-225 *passim*; and McGinn 1998, 200-22, esp. 201-2; see also Tannis M. Guest, "Hadewijch and Minne," in *Studies in the History and Literature of the Netherlands presented to Theodoor Weevers*, ed. P. K. King and P. F. Vincent (Cambridge: Cambridge University Press, 1971), 20ff. Murk-Jansen has observed that *minne* is only the most important of several terms whose ambiguity Hadewijch deliberately exploits to articulate her theological ideas (Murk-Jansen 1992; also Murk-Jansen 1998, esp. 54-57 and 69-71, on *minne*, and 99-100, on punning). Hadewijch's theologically complex concept of *minne* should be compared with that of her older contemporary Beatrijs of Nazareth, which seems similar. Perhaps not surprisingly, Beatrijs uses language of taste and touch in a similar way: see *Van Sieven Manieren van Heiliger Minnen*, ed. H. W. J. Vekeman and J. J. Th. M. Teersteeg (Zutphen: N. V. W. J. Thieme & Cie, 1970), 35ff. (Unfortunately, the *Sieven manieren*, all that survives by Beatrijs herself, is too short to support any strong claims about her use of sensory language.) James A. Wiseman's analysis shows that Ruusbroec uses the term in a closely related way: see "*Minne* in *Die Gheestelike Brulocht*," in Paul Mommaers and Norbert de Paepe, eds., *Jan van Ruusbroec: The Sources, Content and Sequels of his Mysticism* (Leuven: Leuven University Press, 1984), 86-99.

25. Following Barbara Newman, I write "*fine amour*" rather than "courtly love" because the familiar English term has accumulated a tangle of confusing meanings (see Newman, 138, n. 6); moreover, Hadewijch actually uses the term *fine minne* (*PS* 43:29) (which might be translated "refined" or "noble love"). On the Northern French influence, see Reynaert 1994, 208-25.
26. For some specific examples of William's influence, see Paul Verdeyen "De invloed van Willem van Saint-Theirry op Hadewijch en Ruusbroec," *Ons Geestelijk Erf* 51 (1977):3-19; for Hadewijch's use of Richard see J.M. Schalij, "Richard van St. Victor en Hadewijchs 10de Brief," *Tijdschrijft voor Nederlandse Taal- en Letterkunde* 62 (1943), 219-28.
27. Richard of St. Victor, *De quattor gradibus violentiae charitatis*, in *Ives. Épitre à Severin sur la charité. Richard de Saint-Victor. Les quatres degrés de la violent charité*, ed. Gervais Dumeige (Paris: Vrin, 1955). Richard too seems to have drawn on vernacular literature of worldly love.
28. On dynamism and *minne* see de Paepe 1961, 47-49.
29. About *wesen* and *nature*: Hadewijch uses these terms with care and consistency, so I treat them as terms that have distinct fields of meaning with reference to God and holy life. I do not mean to imply that Hadewijch uses them with the same philosophical precision as her scholastic contemporaries did the terms *natura* and *esse* (a caution made by Alaerts with reference to Ruusbroec: "La terminologie 'essentielle' dans *Die gheestlike brulocht*," *Ons Geestelijk Erf* 49 [1975]: 248-330, 337-65).
30. Depth, abyss, ground: e.g., *Letter* 22:8. Sustaining and flowing: *Letter* 22:.251-54, 372-75: "God is outside all and entirely comprised. He is outside

all; for he rests in nothing but the tempestuous nature of his profusely flowing flood, which flows entirely around and over all [things]. . . . Thus God with all the floods of his name is overflowing in all, and around all, and beneath all, and above all; and [he] is comprised in the fruition [enjoyment] of Love." [. . . god buten al es ende al ombegrepen. Hi es buten al: want hine rustet in ghene dinc dan in die druusteghe nature siere vloyender vloedegher vloede, die al omme ende al ouervloyen. . . . Dus es god met alden vloeden van sine name oueruloyende in al ende om al ende onder al ende bouen al ende in ghebrukene van Minnen ombegrepen.]

31. At the end of *Letter* 20 Hadewijch describes the "twelfth nameless hour" of *minne* as what sounds like the inner nature of God's oneness, and she describes *minne*'s essence as her "name." The motif of the "name" or "names" of God is both ancient and common; she herself develops it in *Letter* 22. Her discussion in *Letter* 20 is very similar, and strongly suggests, again, that *minne* is the ground or essential being of God. Hadewijch also uses the pattern "inside-outside, over-under," which echoes the oppositions she uses in *Letter* 22 to talk about God specifically (as distinct from *minne*): ". . . in *minne*'s highest nature . . . [*minne*] first springs out of herself, and she works by herself and she is always so sunken in herself; she makes all satisfaction in her own nature. So she is satisfied in her self: were no one to love *minne,* her name would give her enough lovableness in the honorable [noble] nature of her own self. Her name is her own being within her. Her name is her works outside her. Her name is her crown above her. Her name is her ground under her" (*Letter* 20:123-34). [{In} Minnen ouerste nature . . . daer eerst springhet vte hare seluen Ende si werket met hare seluen, Ende si es soe sinckeleec in hare seluen; si doet al ghenoeghen in hare nature. Si es soe ghenoeghlec in hare seluen: al en Minne nieman de minne, hare name gaue haer minsamheiden ghenoegh in de eersamme nature haers selues. Hare name dat es hare wesen binnen hare. Hare name sijn hare werken buten hare. Haer name es hare crone bouen hare. Hare name dats hare gront onder hare.]
32. Also, in the *Visioenenboek,* she signifies this internal dynamism as a whirling disk in the depth of God: e.g., *Visions* 1:278; 12:1-28
33. *Letter* 12:174; *Letter* 20:99ff; also *PS* 14, *PS* 38:7, and *Visions* 13, esp. ln. 54-57. Hadewijch connects this to the story of Jacob wrestling with and being lamed by the angel (Gen 32:22-32).
34. These are key terms in her writings, and each occur many dozens of times: e.g., *ghebruken*: *Letter* 17:98-101, *PS* 20:40, *Visions* 7:22, *Visions* 8:21, *PM* 1:275; *ghenoeghen*: *Letter* 11:45, *PM* 12:65, *PM* 16:26, *Visions* 7:70-94. On these concepts see e.g., Reynaert 1988, 213-16; on *ghenoeghen,* also Milhaven, 25-34.
35. According to Hadewijch, humanity fell because it failed to meet the reciprocal demands of *minne* that are the internal life of God. This is the main theme of *Letter* 30, esp. ln. 39ff: the Father demands love from the "Unity of the Son and the HS" as they demand it from him. This is "the debt of Unity" paid by Christ who thus ascended to God. Humanity was created in this demand in the

Trinity, but fell because he did not answer the demand. Now we need to make this demand on *minne,* (God), demand his Unity even as we need to pay the debt of *minne* we owe to *minne*. See also *PM* 1.

36. Verwijs, E.-J., and J. Verdam, eds., *Middelnederlandsch woordenboek,* 9 vols. and 2 supplements (The Hague: Nijhoff, 1885-1952), arts. *rueren*; *berueren*; cf. *gherinen*; *gheraken*.
37. Hadewijch uses language of hunger and thirst to teach something very close to Gregory of Nyssa's notion of *epektasis,* the endless pursuit of an endlessly receding God. I learned a good deal about this theme in discussions with Donald Duclow, who has written a paper on the subject: "The Hungers of Hadewijch and Eckhart" *The Journal of Religion* 80 (2000):421-41.
38. She cites it in various renderings, both Latin from the Vulgate ("Dilectus meus mihi et ego illi," *Letter* 13:14) and Dutch ("Du mi al lief ende ic al di," *PS* 34:47). She cites is directly ten times: *Letter* 13:14, 14:38, 19:4, *PM* 16:50, *PS* 12:67, 13:50, 25:9, 27:46, 34:47, 36:92; and echoes it twice: *PS* 38:44 ("ic al di ende du al mi") and *Visions* 1:398-400 ("daer du mi mede best ende ic di" and "salic di ghenoegh wesen ende du mi."
39. *Minne*-nature is not acquired, but inborn in a chosen few: *Visions* 13:195ff. Hadewijch goes on to number them, and in the *List of the Perfect* she gives their names and others'. We see this special status at the start of the passage from *PS* 33:13-28 (cited above, pp. 82-3): those who are born of *minne* share her nature—but this is not everyone, only the "noble." Marguerite Porete has a very similar view. Joanne Maguire has shown that Porete appeals to "nobility" in order to articulate the idea that the soul pre-exists in God: see "Nobility and Annihilation in Marguerite Porete's *Mirror of Simple Souls,*" Ph.D. dissertation, University of Chicago, 1996.
40. Taste death (e.g., Str 24:12, Str 35:13), pain (e.g., *PS* 15:80, *Letter* 1:62), bitterness (e.g., *PS* 17:78), worldly things (e.g., *PS* 11:58).
41. Taste the goodness of God: *PS* 13:64; glory and splendor: *PM* 1:276; hidden word: *PS* 4:17; hidden ways: *PS* 25:35.
42. Taste *minne*: *Letter* 15:78, 18:196, 11:25, 11:44; *PS* 18:25, 20:44, 32:27, 41:8; *PM* 5:2, 6:18; what she is: *PS* 33:34; *minne* nature: *PS* 5:41, 20:40; *Visions* 14:82, 14:163; her being: *PS* 22:32; secret judgments: *Letter* 20:37.
43. It is perhaps not obvious that Hadewijch is speaking of union here; however, she consistently teaches that "(full) enjoyment" (*ghebrukene*), is realized in *minne*. Note also that she links enjoyment and taste; see also *Visions* 8:21 (quoted below, pp. 86-7); and *Visions* 7:22: "Aldus maghicker af segghen: Ic begherde mijns liefs te vollen te ghebrukene ende te bekinnenne ende te ghesmakene in allen vollen ghereke: . . ." [I can say this about it: I desired fully to enjoy my lover/Beloved, and to understand and taste him in all full attainment]). Reynaert closely associates Hadewijch's language of taste with *ghebruken:* Reynaert 1991, 213-16.
44. This is partly an artifact of Middle Dutch usage. As in Modern Dutch, in Middle Dutch the words *smake* (a taste) and *smaken* (to taste) is often used to express the concepts "knowledge" and what we might call "experience." But

I suspect that Dutch scholars sometimes subtly misread Hadewijch's references to taste, because in Middle Dutch this overlap was less fixed than in Modern Dutch, where *smaken* does not seem to be used to refer to the act of tasting with the mouth, but only metaphorically. See E.-J. Verwijs and J. Verdam, eds. *Middelnederlandsch woordenboek* (The Hague: Nijhoff, 1885-1952), art. *smaken* and related terms.

45. Several Cistercian and Victorine writers speak of love and reason as the two eyes of the soul. Hadewijch's discussion in *Letter* 18:80-129 seems to depend on William of St. Thierry's *Natura et dignitate amoris* 8.21-23 (*PL* 184:393-95); and in *Letter* 10 she develops it based on Richard's *Explicatio in Canticum Canticorum* 6 (*PL* 192:422b-23c).
46. This "peacefulness" seems to be that "liberty" or "freedom" (*vrijheit*) from care and concern that a person enjoys when she serves *minne*'s will perfectly; a rest or stability that must still, by definition, partake in the turbulence inherent to *minne*. See, e.g., *Letter* 18:130-73; *PS* 2:77-8.
47. Bernard McGinn, "Love, Knowledge and *Union mystica* in the Western Christian Tradition," in *Mystical Union in Judaism, Christianity and Islam: An Ecumenical Dialogue* (New York: Continuum, 1989), 59-85; see esp. 85.
48. I have translated this sentence ungrammatically because that is the way we have it from the manuscripts. The first clause hangs without a verb: the object of "hevet" is clearly "dat(men)," whose antecedent, grammatically, cannot be "tekenen" but must be "na pleghene van Minnen." So in my translation, "that" refers to "close intercourse of love." Hadewijch often writes with this kind of "broken" syntax, especially in her letters. Modern translations tend to clean it up, obscuring her repetitions and leaps of thought.
49. Agatha Bardoel, in her article "On the Nature of Mystical Experience in the Visions of Hadewijch," *Ons Geestelijk Erf* 66 (1992), 318-40, argues that Hadewijch distinguishes between union *in den gheest* and union *buten den gheest*. However, I think she over-stretches the point. If Hadewijch does make a distinction between these kinds of union, she also undercuts it.
50. I hesitate to call them "rituals" only because I want to refer not only to acts that word immediately brings to mind—formal, "performed" rituals like the Eucharist—but also to acts of piety that the word does not so easily conjure. These would include both the kinds of extraordinary acts of self-sacrifice Bynum analyzes and such everyday acts as care for the sick, instruction, reading and writing. And one of Hadewijch's points is that *all* acts should be done by and for the will of God.
51. This is the plot of the *Book of Visions*. At the very end Hadewijch writes that her narrator, because she is "chosen" can "taste humanity and God in one knowledge" when she is touched, apparently by *minne*. She can do so because she "wholly who our *minne* is': Christ, the Trinity, God (*Visions* 14:161-65; more generally, *Visions* 13 and 14.) This completes the story-line of the book, which is to show how one born of *minne* can realize their latent perfect oneness with God.

Notes to Chapter 5

1. This is amply demonstrated by Marilyn Webster in "Mechthild of Magdeburg's Vocabulary of the Senses" (Ph.D. dissertation, University of Massachusetts Amherst, 1996).
2. Gallus, a canon of St. Victor, later abbot of Vercelli, is a crucial influence on Bonaventure and Rudolf of Biberach. It is from him that they derive the distinctive version of Dionysian apophaticism that is sometimes called "affective Dionysianism," which integrates Dionysius' striking negative theology of "dark" unknowing union with monastic formulations of "affective" union through love and in the *affectus,* on the basis of exegesis of the Song of Songs. They adopt Gallus' terms *apex affectus, apex mentis,* and *synderesis* to label the highest "point" of the soul that knows God nondiscursively and directly while the intellect is blinded and plunged in unknowing. On Gallus and his influence, see McGinn 1998, 78-87.
3. The main passages are as follows (hereafter I will use the abbreviations in parentheses): *Itinerarium mentis in Deum* (*Itin.*) 4; *Breviloquium* (*Brev.*) 5.6; *Commentarius in Libros Sententiarum* (*III Sent.*), dist. 13, dub. 1, dist. 34, art 1; *De reductione artium ad theologiam* (*Red. art.*) 10; *Collationes in Hexaemeron* (*Hexaem.*) III.2, 9, 10. I cite Bonaventure's works by volume and column from the critical edition of the *S. Bonaventurae opera omnia,* 10 vols., (Quaracchi: Collegio S. Bonaventura, 1882-1902) (hereafter cited as *Opera*). Translations are my own, but I have benefitted from several English translations, all based on that edition; in places I have quoted these translations, which I cite at first use.
4. Bonaventure's clearest statements are in *III Sent.*, dist. 13, dub. 1 (*Opera* III, 291-92); dist. 34, art. 1, quest. 1 (*Opera* III, 737b). In *Brev.* 5.6.1 he calls them "*status* delectationum et *usus* spiritualium speculationum, quibus replentur et consolantur spiritus virorum iustorum."
5. Bernard McGinn, "Ascension and Introversion in the *Itinerarium mentis in Deum,*" in *San Bonaventura 1274-1974,* 5 vols. (Grottaferrata and Rome: Collegio S. Bonaventura, 1974) 3:535-52; and McGinn 1998, 105-7.
6. Rahner 1933, 263-99; Hans Urs von Balthasar, "Bonaventure," in *Glory of the Lord* 2:260-362, esp. 309-26 (hereafter cited as von Balthasar). A number of other scholars have also written on Bonaventure's concept of the spiritual senses; there is list through 1968 in the later revision and translation of Rahner's article ("Spiritual Senses in the Middle Ages," in *Theological Investigations* 16, pp. 109-10, n. 23). See esp. R. Carton, *L'expérience mystique de l'illumination interieure chez Roger Bacon* (Paris, 1924), 242-45; J. Bonnefoy, *Le Saint-Esprit et les dons selon S. Bonaventure* (Paris, 1929), esp. 210ff; Ephrem Longpré, "Bonaventure, Saint," *Dictionaire de spiritualité* 1:1826-33; see also Chidester's suggestive account of Bonaventure's language of sight and hearing in *Word and Light,* 115-20.
7. Stephen Fields, S.J., "Balthasar and Rahner on the Spiritual Senses," *Theological Studies* 57 (1996):224-41.

8. Bonaventure specifically introduces Christ as the object of contemplation in this second illuminative stage, in the fourth of the seven *gradus contemplationis*.
9. The phrase is Ephrem Longpré's: "Bonaventure, Saint," *Dictionaire de spiritualité* 1:1831.
10. The whole passage reads as follows: ". . . est homo ad contemplationem idoneus et ad aspectus at amplexus sponsi et sponsae, qui fieri habent secundum *sensus spirituales,* quibus *videtur* Christi sponsi summa plucritudo sub ratione *verbi; gustatur* summa dulcedo sub ratione *sapientiae* comprehendentis utrumque, verbum scilicet et splendorum; *odoratur* summa fragrantia sub ratione *Verbi inspirati* in corde; *astringitur* summa suavitatis sub ratione *Verbi incarnati,* inter nos habitantis [Jn 1:14], corporaliter et reddentis se nobis palpabile, osculabile, amplexabile per ardentissimam caritatem, quae mentem nostram per ecstasim et raptum transire facit *ex hoc mundo ad Patrem* [Jn 13:1]."
11. Translation from *Breviloquium,* trans. Erwin Esser Nemmers (St Louis and London: B. Herder, 1946), 160, slightly altered (hereafter cited as Nemmers). The whole passage reads as follows: "*Sensus* vero *spirituales* dicunt perceptiones mentales circa veritatem contemplandum. Quae quidem contemplatio in Prophetis fuit per *revelationem* quantum ad triplicem visionem, scilicet corporalem, imaginativam et intellectualem; in aliis vero iustis reperitur per *speculationem,* quae incipit a sensu et pervenit ad imaginationem et de imaginatione ed rationem, de ratione ad intellectum, de intellectu ad intelligentiam; de intelligentiae vero ad sapientiam sive notitiam excessivam, quae hic in via incipit, sed consummatur in gloria sempiterna."
12. *Commentarium in Evangelium Lucae* 9.48 (*Opera* 7:231-32) and *Sermo I de Sabbato Sancto* (*Opera* 9:269), where he characterizes the highest contemplation as *gustus, amplexus,* and *requies.* I was made aware of these references and the antecedent in Giles by Bernard McGinn, who cites them in McGinn 1998, 77-78: Giles of Assisi, *Dicta B. Aegidii Assisiensis,* Bibliotheca Franciscana Ascetica Medii Aevi 3 (Quaracchi: Collegium S. Bonaventurae, 1905), cap. 13, pp. 48-9; Thomas Gallus, *De septem gradus contemplationis,* in *S. Bonaventurae Opera Omnia,* ed. A. C. Peltier, 15 vols. (Paris, 1864-1871), 12:183-86. Gallus again alludes to this high ranking for taste in *Commentarium in Canticum Canticorum* (ed. Jeanne Barbet as *Thomas Gallus, Commentaires du Cantique des Cantiques,* Textes Philosophiques du Moyen Age 14 [Paris: 1967]); passage quoted below, p. 158, n. 16.
13. It might be objected here that the object of contemplation and union never really changes because Bonaventure presents all the soul's various objects, from more corporeal to more spiritual, as aspects of the Word. The Father created all things through the Word, and strictly speaking the Word remains our only access to him. However, Bonaventure makes distinctions. It is clear that in the highest stages of contemplation and union the soul knows and is one with the more divine and spiritual aspects of the Word, and ultimately with its transcendent aspect. Bonaventure specifies that in the unitive stages the soul con-

templates "the eternal things of God, . . . [his] essential attributes, and the properties of the Persons" of the Trinity. It may be objected, again, that Bonaventure specifies that ecstasy is contemplation of *the God-man,* whose dual nature manifests the "coincidence of opposites" that best approximates God's the incomprehensible transcendence. But note that his emphasis is less on the historical Christ as such than on the transcendent attributes of God that are manifested by the paradox of the Word incarnate. This emphasis makes more sense when we consider that Bonaventure uses Dionysian language to characterize ecstatic union with God as a paradoxical "superluminous darkness." He wants to emphasize that God in himself transcends what we are and how we can think, and in ecstasy the highest and most spiritual part of the soul knows—or better, inhabits—this unknowability. Ecstasy is paradoxical precisely because in ecstasy the soul's object is *not* the mediating immanent Word, it is God's and/or the Word's incomprehensible transcendence.

14. *Quaestiones disputata de scientia Christi,* q. 7, ad 19-21 (*Opera* 5:43a): ". . . excessus est ultimus modus cognoscendi et nobilissimus" (quoted in McGinn 1998, 111 and n. 215). Bonaventure's ambiguity has precipitated a long-standing scholarly debate about whether he gives primacy to "love" or "knowledge." McGinn's interpretation, which I follow here, cites the disputed passages and all the main scholarly opinions (which include those of Rahner, Longpré, and Gilson): see McGinn 1998, 110-11 and the accompanying notes.
15. McGinn has argued for this interpretation: "What becomes present in the *apex* [*affectus*] cannot be known and expressed. . . . But the *apex* still bears some kind of relation to knowledge and to what can be expressed in language . . . because it draws up into itself all the preparatory cognitive operations that are part of the journey into God just at the moment it leaps beyond them. . . ." McGinn 1998, 111.
16. He quotes Gallus in support in *De septem itineribus aeternitatis,* in *S. Bonaventurae Opera Omnia,* 15 vols., ed. A.C. Peltier (Paris: 1864-1871), vol. 8, pp. 393-482; reprinted as *Rudolf von Biberach, De septem intineribus aeternitatis,* with an introduction and corrections by Margot Schmidt, Mystik in Geschichte und Gegenwart, Texte und Untersuchungen I,1 (Stuttgart-Bad Cannstatt: 1985), 6.6 (hereafter cited as *Sept. itin.,* by chapter and section; all translations are my own). The quotation is as follows: "Nihil est quod ita gratam mentibus praestet refectionem, sicut panis supersubstantialis. Haec enim refectio non fit per speculum, sed per divinae dulcedinis experientiam, juxta quod gustus et tactus non excercentur per speculum, sicut visus." It is from Thomas Gallus, *Commentarius in Canticum Canticorum* [*Tertius*] 1.C, which can be found in *Commentaires du Cantique des Cantiques,* ed. Jeanne Barbet [Paris, Vrin, 1967], "Troisième Commentaire," 124. I have not searched Gallus' writings thoroughly, but I note that near the end of the Prologue to his "Second" commentary on the Song of Songs, he alludes to the immediacy of touch when speaking of the highest hierarchy of mind: the "ecstasies and excesses" that exceed understanding and sight he calls "embraces" and distinguishes from "knowledge in a mirror," as by sight (Barbet, 67).

17. The book is almost entirely a collection of quotations from other authors, which are arranged into "distinctions" and "articles" in imitation of a scholastic treatise. The book seems to have been used (and may have been intended) as a handbook for spiritual directors; on this genre see Margot Schmidt (citation below, in this note), and McGinn 1988, 116-20. There is some scholarship on Rudolf's account of the spiritual senses: Rahner mentions it (Rahner 1933, 292-93), as does von Balthasar (Balthasar 1:373) and Canévet; Margot Schmidt gives a more extensive summary with a few analytical remarks in her introduction to Rudolf von Biberach, *Die siben strassen zu got. Die hochalemannische Übertragung nach der Handschrift Einsiedeln 278,* ed. with an introduction by Margot Schmidt (Quaracchi, Florence, Collegii S. Bonaventurae, 1969), 219-25 (hereafter cited as Schmidt 1969); and again in her clear and concise article "Rodolphe de Biberach," *Dictionaire de spiritualité* 13: col. 846-50, esp. 848-9 (hereafter cited as Schmidt, *DS*).
18. Here again we see the influence of Gallus' "affective Dionysianism," which in the *Seven Roads* Rudolf is much concerned to promote. He quotes Gallus often and at length (although rarely by name), and the fact that Rudolf's book circulated widely under Bonaventure's name certainly helped secure a wide influence for Gallus' ideas and key terms.
19. This is not a direct quotation of Richard, but Rudolf seems to have in mind *Benjamin major* [*De arce mystica*] 9.15, which he cites just after this.
20. On Rudolf's ideas about the Eucharist, see, in addition to Schmidt 1969, Introduction; Schmidt, *DS;* and E. Longpré, "L'Eucharistie et l'union mystique selon la spiritualité franciscaine," *Revue d'ascetique et de mystique* 25 (1945):306-33; on Rudolf, 322-27.
21. Rudolf's main reason for placing sight first and lowest among the spiritual senses is certainly to emphasize that intellect and rational knowing, so closely associated with sight, are outstripped by affective, "experiential" knowing through love.
22. Ruusbroec's language of taste is equally significant, but to spare us the sight of a well-flogged dead horse I pass over it. I note only that Ruusbroec, very much like the other theologians I discuss, frequently uses language of taste to articulate immediate union with God, which gives knowledge exceeding the capacity of intellect and reason. Ruusbroec also, of course, often uses language of sight and light to discuss the contemplative's bond and ultimate union with God. In this he demonstrates his knowledge of explicit theological speculation on *contemplatio* and *visio Dei*. He is also well-known for some synaesthetic passages about union with God, especially in the final pages of *The Spiritual Espousals* and *The Sparkling Stone.* I cite Ruusbroec's works from the new *Opera omnia* being published in the series Corpus Christianorum Continuatio Medievalis (CCCM) (Turnhout: Brepols, 1988-), 10 volumes projected (hereafter cited as *Opera*); CCCM 103 contains *The Spiritual Espousals* (*Die Gheestelike Brulocht*; hereafter *Espousals*), ed. J. Alaerts with English translation by Helen Rolfson (1988); CCCM 110 contains *The Sparkling Stone* (*Vanden Blinkenden Steen*; hereafter *Stone*), ed. G. de Baere, Th. Mertens and

H. Noë, with English translation by A. Lefevre (1991). The exception is *The Mirror of Eternal Blessedness* (*Een Spieghel der Eeuwigher Salicheit*; hereafter cited as *Mirror*), which has not yet appeared in this new edition; I cite it from Jan Ruusbroec, *Werken*, 2d ed., revised, ed. by the Ruusbroecgenootschap, 4 vols, (Tielt: Lannoo, 1944-48), vol. 3, ed. L. Reypens, S.J. and M. Schurmans S.J., 127-219 (hereafter cited as *Werken*). I cite *Espousals* and *Stone* according to the new line enumeration in the *Opera*, but also give a cross-reference to the page in the *Werken*, since that edition is probably still more widely available. Some translations are mine, as noted; others I quote from the generally excellent new translations in the *Opera omnia*, and in the case of the *Mirror* from *John Ruusbroec: The Spiritual Espousals and Other Works*, trans. James A. Wiseman (New York and Mahwah, NJ: Paulist Press, 1985) (hereafter cited as Wiseman).

23. This is a major theme in many of his works: see e.g., *Espousals* b 2294-584 (*Werken*, 228-38); *Stone* 191- 476 (*Werken*, 11-22); *Mirror* 2.B, groups 5-7 (*Werken*, 188-95).
24. Leonce Reypens long ago drew attention to what he called Ruusbroec's concept of "divine touch" and argued that it is unequivocally a doctrine of grace: *idem*, "Ruusbroec-Studien. Het mystieke 'gherinen,'" *Ons Geestelijk Erf* 12 (1938):15-86 (hereafter cited as Reypens). Ruusbroec does not directly cite or paraphrase Bonaventure, but it is hard to doubt that he knew something of Bonaventure's ideas. We know Ruusbroec read at least some of Hadewijch's works: he mentions her twice, and there are a number of passages that have been identified as paraphrases. The two main discussions are Stephanus Axters, O.P., "Hadewijch as voorloopster van de zalige Jan van Ruusbroec," in *Dr. L. Reypens-album*, ed. Albert Ampe, S.J. (Antwerp: Ruusbroecgenootschaap, 1964), 57-74, and Joris Reynaert, "Ruusbroec en Hadewijch," *Ons Geestelijk Erf* 55 (1982):193-233. More generally, see also G. Epiney-Burgard, "L'influence des béguines sur Ruusbroec," in Paul Mommaers and Norbert de Paepe, eds., *Jan van Ruusbroec: The Sources, Content, and Sequels of his Mysticism* (Leuven: Leuven University Press, 1984), 68-85. With specific reference to sensory language, Reypens notes briefly that there are antecedents for Ruusbroec's use of tactile language in Hadewijch and Beatrijs of Nazareth, and suggests direct influence (Reypens, 15, nn. 2 and 3). On the other hand, Reynaert notes that such antecedents do not prove influence, because sensory language is commonplace in Middle Dutch texts (as it is in writing in Latin and other late-medieval European languages). He argues that between Ruusbroec and Hadewijch there is a "connection," but not direct "influence," because in Ruusbroec's writings the word *gherinen* is a "technical term," while in Hadewijch's it has only a "general and vague meaning." I would suggest that Reynaert's claim labels less a difference in *concepts* than a difference in *genres*. In any case, whether or not Ruusbroec derived his concept of *gherinen* from her, it shows remarkable similarities to hers, and—no coincidence—so does the exemplarist, Trinitarian theology that it supports and is supported by.

25. Ruusbroec uses all these phrases frequently: see, e.g., *Espousals* c 36 (*Werken,* 240); b 1481, b 1582 (*Werken,* 197, 201).
26. E.g., *Espousals* b 1569, b 1572 (*Werken,* 200).
27. Reypens gives a long list of citations; see, e.g., the major passage on touch in *Espousals*, which has a dozen examples: b 1462-1602, *passim* (*Werken,* 196-202).
28. *Espousals* b 1588-91; *Werken,* 201 (cf. Hadewijch, *Mgld.* 16, quoted above, p. 97); more extensively, *Mirror* 2.B, third point (*Werken,* 158-61). We have seen that Bernard too flirts with the idea of reciprocity, especially when he articulates union as God and the soul eating each other, and each engulfing the other. Bernard certainly sees its heterodox implications, but doesn't worry about them too much, probably because his audience of monks would not likely misunderstand him on this point. Hadewijch also seems to see the implications, but doesn't seem to consider them heterodox and lets them stand.
29. This point has been emphasized by Paul Henry, "La mystique trinitaire de bienheureux Jean Ruusbroec" *Recherches de science religieuse* 39-40 (1951-52):335-68, 41 (1953):51-75, esp. 62; and by Louis Dupré, *The Common Life: The Origins of Trinitarian Mysticism and its Development by Jan van Ruusbroec* (New York: Crossroad, 1984) (hereafter cited as Dupré).
30. He discusses spiritual touch again, without deviating significantly from his presentation in *Espousals* and *Stone*, in his later work *The Seven Enclosures* (*Vanden Seven Sloten*), ed. G. de Baere, in *Opera Omnia*, Corpus Christianorum Continuatio Medievalis vol. 102 (Turnhout: Brepols, 1984); see esp. lines 636-45.
31. Ruusbroec, therefore, teaches that union of spirit first with then without intermediary (our topic here) is complemented and perfected by union without difference. For in that higher or more immediate union a person lives the *full* living economy of God, which is both active (*werkelijc*), and idle and empty (*ledich*); both perfect and poised in itself, and fully out-flowing to others. Again, Ruusbroec presents these two forms of union as steps in a sequence, but emphasizes that they simultaneous aspects of one dialectical union. On this major theme in Ruusbroec's theology of union, see Dupré.

Notes to Epilogue

1. On Ruusbroec in particular, see Helmut Hatzfeld, "The Influence of Ramon Lull and Jan van Ruysbroeck on the Language of the Spanish Mystics," *Traditio* 4 (1946):337-97; on "mystical touch," see esp. 382-84. There is a huge literature on Ignatius' ideas about how to use and control the bodily senses for spiritual ends; Canévet surveys it and gives a short bibliography. As for Teresa and John, my friend and colleague Edward Howells has examined their ideas about theological anthropology and touches on their ideas about the spiritual senses in "Mystical Consciousness and the 'Mystical Self' in Teresa of Avila and John of the Cross" (Ph.D. dissertation, University of Chicago, 1999).

2. Here again I refer the reader to Sharf, who clearly and unsentimentally outlines the complications and implications of the term "experience" with reference to the field of religious studies.

Select Bibliography

ANCIENT AND MEDIEVAL LITERATURE

Albert the Great. *Opera omnia*. Edited by A. Borgnet. Paris: Vivés, 1890-1895.

Alexander of Hales. *Summa theologicae*. Edited by Bernardin Klumper. Quaracchi: Typographia Collegii S. Bonaventurae, 1928.

Aristotle. *The Works of Aristotle Translated into English*. 12 vols. Edited by J. A. Smith and W. D. Ross. Oxford: 1908-1952.

Augustine. *Confessions*. 2 vols. Edited by W. H. D. Rouse, with the translation of W. Watts (London: 1631). Loeb Classical Library. London: William Heinemann and Cambridge, MA: Harvard University Press, 1912; reprint, 1960.

———. *The City of God*. Translated by Marcus Dods, George Wilson, and J. J. Smith. Modern Library. New York: Random House, 1950.

Beatrijs of Nazareth. *Van Sieven Manieren van Heiliger Minnen*. Edited by H. W. J. Vekeman and J. J. Th. M. Teersteeg. Zutphen: N. V. W. J. Thieme & Cie, 1970.

Bernard of Clairvaux. *Sancti Bernardi opera omnia*. 8 vols. Edited by Jean Leclercq, C. H. Talbot, and H. M. Rochais. Rome: 1957-1980.

———. *On the Song of Songs*. 4 vols. Cistercian Fathers 4, 7, 31, 40. Translated by Kilian Walsh, O.C.S.O., and Irene M. Edmonds. Kalamazoo: Cistercian Publications, 1971-1983.

———. *Five Books on Consideration: Advice to a Pope*. Cistercian Fathers 37. Translated by John Anderson and Elizabeth T. Kennan. Kalamazoo, MI: Cistercian Publications, 1976.

———. *Sermons on Conversion: On Conversion, a Sermon to Clerics, and Lenten Sermons on the Pslam 'He Who Dwells.'* Cistercian Fathers 25. Translated by Marie-Bernard Saïd, OSB. Kalamazoo, MI: Cistercian Publications, 1981.

———. *Selected Works*. Translated by G. R. Evans. New York and Mahwah, NJ: Paulist Press, 1987.

———. *On Loving God, with an Analytical Commentary by Emero Steigman*. [Translated by Robert Walton.] Cistercian Fathers 13b. Kalamazoo, MI: Cistercian Publications, 1995.

Biblia sacra iuxta vulgatam versionem. Edited by Bonifatus Fischer, OSB, *et al*. Stuttgart: Deutsche Bibelgesellschaft, 1983.

Bonaventure. *Opera Omnia*. 10 vols. Quaracchi: Collegio S. Bonaventura, 1882-1902.

———. *Breviloquium*. Translated by Erwin Esser Nemmers. St Louis and London: B. Herder, 1946.

———. *The Mind's Road to God*. Translated by George Boas. Library of Liberal Arts. New York: Macmillan and London: Collier, 1953.

Cicero, Marcus Tullius. *De natura deorum*. Edited and translated by H. Rackham. Loeb Classical Library. London: William Heinemann and Cambridge, MA: Harvard University Press, 1933.

Hadewijch. *Hadewijch: Visioenen*. 2 vols. Edited by Jozef van Mierlo, S.J. Leuvense Studiën en Tekstuitgaven. Louvain: Vlaamsche Boekanhalle, 1924, 1925.

———. *Hadewijch: Strofische Gedichten*. 2 vols. Edited by Jozef van Mierlo, S.J. Leuvense Studiën en Tekstuitgaven, 2 vols. Antwerp: Standaard, 1942.

———. *Hadewijch: Brieven*. 2 vols. Edited by Jozef van Mierlo, S.J. Leuvense Studiën en Tekstuitgaven. Antwerp: Standaard, 1947.

———. *Hadewijch: Mengeldichten*. Edited by Jozef van Mierlo, S.J. Leuvense Studiën en Tekstuigaven. Antwerp: Standaard, 1952.

———. *Écrits mystiques des Béguines: Hadewijch d'Anvers* [a selection of *Strophische Gedichten* and *Mengeldichten*]. Translated (into French) with notes by Fr. J[ean].-B[aptiste]. M. P[orion]. Paris: Éditions du Seuil, 1954.

———. *Hadewijch. Strofische Gedichten*. Edited and translated (into Modern Dutch) by E. Rombauts and Norbert de Paepe. Zwolle: W. E. J. Tjeenk Willink, 1961.

———. "Hadewijch of Antwerp: Letters [1-20]." In *Medieval Netherlands Religious Literature*, edited and translated by Eric Colledge. Leiden: Sythoff; London: Hainemann, and New York: London House and Maxwell, 1965:31-67.

———. *De Wetten van de Minne: met de tekst van Hedewijchs 45 Strofische Gedichten volgens HS. 385 II* [ms R] *van het Ruusbroecgenootschap*. 2 vols. Edited by Joseph Alaerts, S.J. Bonheiden: Zusters Benedictinessen Priorij Bethlehem [1977].

———. *De visionen van Hadewijch*. 2 vols. Edited and translated (into Modern Dutch) with commentary by Paul Mommaers. Nijmegen: 1979.

———. *Het visioenenboek van Hadewijch: uitg. naar Handschrift 941 van de Bibliotheek der Rijksuniversitaet te Gent*. Edited and translated (into Modern Dutch) with commentary by H. W. J. Vekeman. Nijmegen: Dekker Van de Vegt; Brugge: Orion, 1980.

———. *Hadewijch: Complete Works*. Translated by by Mother Columba Hart, OSC. New York: Paulist Press, 1984.

———. *Amour est tout: poèmes strophiques*. Translated (into French) by Rose Vande Plas. Paris: Téqui, 1984.

———. *Visions*. Translated (into French) by Fr. J[ean].-B[aptiste]. M. P[orion]. Paris: Guibert Fox, 1987.

———. *De visionen van Hadewijch*. Translated (into Modern Dutch) by Imme Dros, edited by Frank Willaert. Amsterdam: Uitgeverij Prometheus/Bert Bakker, 1996.

———. *Poetry of Hadewijch [Strofische Gedichten]*. Translated by Marieke J. E. H. T. van Baest. Leuven: Peeters, 1998.

Hugh of St. Victor. *Soliloquium de arrha anima*. In *Patrologiae cursus completus series latina*, edited by J. P. Migne. Paris: 1844-64. 175:951-63.

Origen. *The Commentary of Origen on St. John's Gospel*, 2 vols. Edited and translated by A. E. Brooke. Cambridge: Cambridge University Press, 1896.

———. *Against Celsus*. Translated by Henry Chadwick. Cambridge: Cambridge University Press, 1953.

———. *The Song of Songs, Commentary and Homilies*. Translated by R. P. Lawson. Ancient Christian Writers 26. New York and Ramsey, NJ: Paulist Press, 1956.

———. *Entretien d'Origène avec Héraclide*. Edited and translated (into French) by Jean Scherer. Sources Chrétiennes 67. Paris: Cerf, 1960.

———. *Contre Celse*. 5 vols. Edited and translated (into French) by M. Borret. Sources Chrétiennes 132, 136, 147, 150, 227. Paris: Cerf, 1967-1976.

———. *On First Principles*. Translated by G. W. Butterworth. Gloucester, MA: Peter Smith, 1973.

———. *Traité des principes*. 5 vols. Edited and translated (into French) by Henri Crouzel and M. Simonetti. Sources Chrétiennes 252, 253, 268, 269, 312. Paris: Cerf, 1978-1984.

———. *Commentaire sur le Cantique de Cantiques*. 2 vols. Edited and translated (into French) by L. Brésard, H. Crouzel, and M. Borret. Sources Chrétiennes 375, 376. Paris: Cerf, 1991, 1992.

———. *Treatise on the Passover and Dialogue with Heraclides*. Translated by Robert J. Daly. Ancient Christian Writers 54. New York : Paulist Press, 1992.

———. *Der Kommentar zum Evangelium nach Mattäus*. Edited and translated (into German) with notes by Hermann J. Vogt. Stuttgart: Hiersemann, 1993.

Plato. *The Works of Plato*. Translated by Benjamin Jowett. Selected and edited by Irwin Edman. Modern Library. New York: Random House, 1928.

Richard of St Victor. *De quattor gradibus violentiae charitatis*. In *Ives. Épitre à Severin sur la charité. Richard de Saint-Victor. Les quatres degrés de la violent charité*. Edited by Gervais Dumeige. Paris: Vrin, 1955.

———. *Richard of St. Victor: Selected Writings on Contemplation*. Translated by Clare Kirchberger. New York: Harper, n.d.

Rudolf of Biberach. *De septem itineribus aeternitatis*. In *S. Bonaventurae Opera Omnia*, 8:393-482. Edited by A.C. Peltier. Paris: 1866. Reprinted as *Rudolf von Biberach, De septem intineribus aeternitatis*, with an introduction and corrections by Margot Schmidt. Mystik in Geschichte und Gegenwart. Texte und Untersuchungen I,1. Stuttgart-Bad Cannstatt: Frommann-Holzboog, 1985.

Ruusbroec, Jan. *Die Gheestelike Brulocht*. Edited by J. Alaerts with English translation by Helen Rolfson and Latin translation of L. Surius (Cologne, 1552). Turnhout: Brepols and Tielt: Lanoo, 1988. Corpus Christianorum Continuatio Medievalis 103. *Jan van Ruusbroec opera omnia*. Edited by G. de Baere.

———. *Vanden Blinkenden Steen, Vanden Vier Becoringhen, Vanden Kerstenen Ghelove, Breiven*. Edited by G. de Baere, Th. Mertens and H. Noë, with English translation by A. Lefevre and Latin translation of L. Surius (Cologne, 1552). Turnhout: Brepols and Tielt: Lanoo, 1991. Corpus Christianorum Continuatio Medievalis 110. *Jan van Ruusbroec opera omnia*, vol 10. Edited by G. de Baere.

———. *Werken*. Edited by the Ruusbroecgenootschap. 2d ed., revised. 4 vols. Tielt: Lannoo, 1944-1948.

———. *John Ruusbroec: The Spiritual Espousals and other Works*. Translated by James A. Wiseman. New York and Mahwah, NJ: Paulist Press, 1985.

Septuaginta, id est Vetus Testamentum Graece iuxta LXX interpres. Stuttgart: Deutsche Bibelgesellscahft, 1980.

Thomas Aquinas. *Opera omnia*. 26 vols. Edited by Vernon J. Bourke. New York: Musurgia Publishers, 1948-1950; reprint of the "Parma" edition, Parma: 1852-1873.

Thomas Gallus. *Commentaires du Cantique des Cantiques*. Edited by Jeanne Barbet. Paris: Vrin, 1967.

William of Auvergne. *Opera Omnia [apud A. Pralard] 1674*. 2 vols. Facsimile edition. Frankfurt am Main: Minerva, 1963.

William of Auxerre. *Summa aurea* [*Summa super quatuor libros sententiarum*]. 4 vols. Edited by Jean Ribaillier. Paris: Éditions du Centre national de la recherche scientifique and Rome: Editiones Collegii S. Bonaventurae ad Claras Aquas, 1980-1987.

William of St. Thierry. *De natura et dignitate amoris*. In *Patrologiae cursus completus series latina,* edited by J. P. Migne. Paris: 1844-64. 184:379-408.

———. *Speculum fidei*. In *Patrologiae cursus completus series latina,* edited by J. P. Migne. Paris: 1844-1864:180:365-87.

MODERN STUDIES

Alexandre, Monique. "Le statut des questions concernant la matiére dans le *Peri Archon*." In *Origeniana*, edited by H. Crouzel et al. Bari: Instituto di Letteratura Christiana Antica, 1975:63-81.

Axters, Stephanus, O.P. *The Spirituality of the Low Countries*. Translated by Donald Attwater. London: Blackfriars, 1954.

———. "Hadewijch as voorloopster van de zalige Jan van Ruusbroec." In *Dr. L. Reypens-album*, edited by A. Ampe. Antwerp: Ruusbroecgenootschaap, 1964:57-74.

Balas, David. "The Idea of Participation in the Structure of Origen's Thought." In *Origeniana*, edited by H. Crouzel et al. Bari: Instituto di Letteratura Christiana Antica, 1975:257-75.

Balthasar, Hans Urs von. *The Glory of the Lord: A Theological Aesthetics*. 7 vols. Edited by Joseph Fessio, S.J., and John Richies. San Francisco: Ignatius Press and New York: Crossroad, 1982-1989.

Bardoel, Agatha Anna. "On the Nature of Mystical Experience in the Visions of Hadewijch." *Ons Geestelijk Erf* 66 (1992):318-40.

Bell, David N. *Image and Likeness: The Augustinian Spirituality of William of St. Thierry*. Cistercian Studies 78. Kalamazoo, MI: Cistercan Publications, 1984.

Blanpain, Jacques. "Langage mystique, expression du désir dans les Sermons sur le Cantique des cantiques de Bernard de Clairvaux." *Collectanea Cisterciensia* 36 (1974):45-68, 225-47; 37 (1975): 145-66.

Bonowitz, J. "The Role of Experience in the Spiritual Life." In *La dottrina della vita spirituale nelle opera di San Bernardo di Clairvaux: Analecta Cisterciana* 46 (1990):321-25.

Bostock, D.G. "Quality and Corporeity in Origen." In *Origeniana Secunda*, edited by H. Crouzel and A. Quacquarelli. Edizioni dell Ataneo, 1980:323-37

Bynum, Caroline Walker. *Holy Feast and Holy Fast: The Significance of Food to Medieval Women*. Berkeley and Los Angeles: University of California Press, 1988.

———. *Fragmentation and Redemption*. Cambridge, MA: Zone, 1991.

———. *The Resurrection of the Body in Western Christianity, 200-1336.* New York: Columbia University Press, 1995.

Canévet, Mariette. "Sens Spirituel." *Dictionaire de spiritualité* 14, col. 598-617.

Casey, Michael. *Athirst for God: Spiritual Desire in Bernard of Clairvaux's Sermons on the Song of Songs.* Kalamazoo, MI: Cistercian Publications, 1988.

Chadwick, Henry. "Origen, Celsus, and the Resurrection of the Body," *Harvard Theological Review* 41 (1948):83-102.

Chidester, David. "Symbolism and the Senses in Saint Augustine." *Religion* 14 (1984):31-51.

———. *Word and Light: Seeing, Hearing and Religious Discourse.* Urbana and Chicago: University of Illinois Press, 1992.

Classen, Constance. *Worlds of Sense: Exploring the Senses in History and Across Cultures.* London and New York: Routledge, 1993.

Cousins, Ewert. "The Humanity and The Passion of Christ." In *Christian Spirituality II: High Middle Ages and Reformation,* edited by Jill Raitt, in collaboration with Bernard McGinn and John Meyendorff. New York: Crossroad, 1988:375-91

Cox, Patricia. "'Pleasure of the Text, Text of Pleasure': Origen's *Commentary on the Song of Songs.*" *Journal of the American Academy of Religion* 54 (1986):241-51.

Crouzel, Henri. *Origéne et la "connaissance mystique."* Paris and Bruges: 1961.

———. *Bibliographie critique d'Origène*, The Hague: Nijhoff, 1971, and *Supplément*, 1982.

———. "Les critiques addressés par Methode et ses contemporains à la doctrine origenienne du corps résuscité." *Gregorianum* 53 (1972):649-716.

———. "La doctrine origenienne du corps résuscité." *Bulletin de littérature ecclesiastique* 31 (1980):175-200, 241-66.

———. *Origen.* Translated by A. S. Worrall. San Francisco: Harper and Row, 1989. Translation of *Origène.* Paris: Éditions Lethielleux and Namur: Culture et Vérité, 1985.

Crouzel, Henri, et al., eds. *Origeniana.* Bari: Instituto di Letteratura Christiana Antica, 1975.

Damasio, Antonio. *Descartes' Error: Emotion, Reason and the Human Brain.* New York: Avon Books, 1994.

Danesi, M. "Thinking is Seeing: Visual Metaphors and the Nature of Abstract Thought." *Semiotica* 80 (1990):221-27.

Deblaere, Albert. "Témoinage mystique chrétien," *Studia Missionalia* 26 (1977):127.

DeGanck, Roger. *Beatrice of Nazareth and the Thirteenth-Century* mulieres religiosae *of the Low Countries*, 3 vols. Cistercian Fathers 50, Cistercian Studies 121 and 122. Kalamazoo, MI: Cistercian Publications, 1991.

Dechow, Jon F. *Dogma and Mysticism in Early Christianity: Epiphanius of Cyprus and the Legacy of Origen.* Macon, Georgia: Mercer University Press, 1988.

Deroy, Jean Prosper Theodorus. *Bernardus en Origenes: Enkele opmerkingen over de invloed van Origenes op Sint Bernardus*. Haarlem: De Toorts, 1963.

Dillon, John. "*Aesthésis Noété*: A Doctrine of Spiritual Senses in Origen and Plotinus." In *Judaica et Hellenica: Hommage à V. Nikiprowetzky*, edited by A. Caquot, M. Hadas-Lebel and J. Riaud. Louvain and Paris: Peters, 1986:443-55.

Doyère, P. "Sainte Gertrude et les sens spirituels." *Revue d'ascetique et de mystique* 36 (1960):429-48.

Dupré, Louis. *The Common Life: The Origins of Trinitarian Mysticism and Its Development by Jan van Ruusbroec.* New York: Crossroad, 1984.

Elorduy, E. "El influjo estóico en Origenes." In *Origeniana*, edited by H. Crouzel et al. Bari: Instituto di Letteratura Christiana Antica, 1975:277-88.

Epiney-Burgard, Georgette. "L'influence des béguines sur Ruusbroec." In *Jan van Ruusbroec: The Sources, Content and Sequels of his Mysticism,* edited by Paul Mommaers and Norbert de Paepe, 68-85. Leuven: Leuven University Press, 1984.

Ewer, Mary Anita. *A Survey of Mystical Symbolism.* London: S.P.C.K. and New York: Macmillan, 1933.

Fernandez, James, ed. *Beyond Metaphor: The Theory of Tropes in Anthropology.* Stanford, CA: Stanford U. Press, 1991.

Fields, Stephen, S.J. "Balthasar and Rahner on the Spiritual Senses." *Theological Studies* 57 (1996):224-41.

Fraigneau-Julien, B. *Les sens spirituels et la vision de Dieu chez Syméon le Nouveau Théologien*. Théologie Historique 67. Paris: 1985.

Gilson, Etienne. *The Philosophy of St. Bonaventure.* Translated by Illtyd Trethowan and F.J. Sheed. London: Sheed and Ward, 1938; reprint, Paterson, NJ: St. Anthony Guild Press, 1965.

———. *History of Christian Philosohopy in the Middle Ages.* London: Sheed and Ward, 1980.

———. *The Mystical Theology of St. Bernard.* Translated by A.H.C. Downes. Kalamazoo: Cistercian Publications, 1990.

Grant, Robert. "Marcion and the Critical Method." In *From Jesus to Paul*, edited by Peter Richardson and John C. Hurd. Waterloo, ON: Wilfrid Laurier University Press, 1984:207-15.

———. *The Letter and the Spirit*. New York: Macmillan, 1957.

Greenspan, Kate. "Autohagiography and Medieval Women's Spiritual Autobiography." In *Gender and Text in the Later Middle Ages*, edited by Jane Chance. Gainesville, FL: University Press of Florida, 1996:216-36.

Grundmann, Herbert. *Religious Movements in the Middle Ages*. Translated by Steven Rowan. Notre Dame: University of Notre Dame Press, 1996.

Guest, Tannis M. "Hadewijch and Minne." In *Studies in the History and Literature of the Netherlands presented to Theodoor Weevers*, edited by P. K. King and P. F. Vincent. Cambridge: Cambridge University Press, 1971.

Hale, Rosemary Drage. "'Taste and See, for God is Sweet': Sensory Perception and Memory in Medieval Christian Mystical Experience." In *Vox Mystica: Essays on Medieval Mysticism*, edited by Anne C. Bartlett et al. Cambridge: Brewer, 1995:3-14.

Hamburger, Jeffrey. *The Rothschild Canticles: Art and Mysticism in Flanders and the Rhineland circa 1300*. New Haven: Yale, 1990.

Hanson, R. P. C. *Allegory and Event: A Study of the Sources and Significance of Origen's Interpretation of Scripture*. London: SCM Press, 1959.

Harl, Margeurite. *Origène et la fonction révélatrice du Verbe incarné*. Paris: Éditions du Seuil, 1958.

———. "La 'bouche' et le 'coeur' de l'apôtre: deux images bibliques du 'sens divin' de l'homme ('Proverbes' 2,5) chez Origène." In *Forma Futuri: Studi in Onore del Cardinale Michele Pellegrino*. Turin: Bottega d'Erasmo, 1975:17-42.

Helmut Hatzfeld. "The Influence of Ramon Lull and Jan van Ruysbroeck on the Language of the Spanish Mystics." *Traditio* 4 (1946):337-97.

Hennessy, Lawrence. "A Philosophical Issue in Origen's Eschatology: The Three Senses of Incorporeality." In *Origeniana Quinta*, edited by Robert J. Daly. Leuven and Paris: Peeters, 1992:373-80.

Henry, Paul. "La mystique trinitaire de bienheureux Jean Ruusbroec." *Recherches de science religieuse* 39-40 (1951-52):335-68, 41 (1953):51-75.

Hollywood, Amy. *The Soul as Virgin Wife: Mechthild of Magdeburg, Marguerite Porete and Meister Eckhart*. Notre Dame and London: University of Notre Dame Press, 1995.

Howells, Edward. "Mystical Consciousness and the 'Mystical Self' in Teresa of Avila and John of the Cross." Ph.D. dissertation, University of Chicago, 1999.

Howes, D., ed. *The Varieties of Sensory Experience: A Sourcebook in the Anthropology of the Senses*. Toronto: University of Toronto Press, 1991.

Hume, David. "Of Miracles." In: *Writings on Religion,* edited by Anthony Flew. La Salle, IL: Open Court Press, 1992:63-88.

Jackson, Michael. "Thinking Through the Body: An Essay on Understanding Metaphor." *Social Analysis, Journal of Cultural and Social Practice* 14 (Dec 1983):127-49.

Jantzen, Grace M. *Power, Gender and Christian Mysticism*. Cambridge, UK: Cambridge University Press, 1995.

Jay, Martin. *Downcast Eyes: The Denigration of Vision in Twnetieth-Century French Thought*. Berkeley: University of California Press, 1993.

Johnson, Mark. *The Body in the Mind: The Bodily Basis of Metaphor, Imagination, and Reason*. Chicago: University of Chicago Press, 1987.

Jonas, Hans. "The Nobility of Sight: A Study in the Phenomenology of the Senses." In *The Phenomenon of Life: Toward a Philosophical Biology*. Chicago: University of Chicago Press, 1982:135-56.

Katz, Steven, ed. *Mysticism and Philosophical Analysis*. New York: Oxford University Press, 1978.

Kay, Sarah. *Subjectivity in Troubadour Poetry*. Cambridge and New York: Cambridge University Press, 1990.

Kieckhefer, Richard. *Unquiet Souls: Fourteenth-Century Saints and Their Religious Milieu*. Chicago: University of Chicago Press, 1984.

Köpf, Ulrich. *Religiöse Erfahrung in der Theologie Bernhards von Clairvaux*. Tübingen: Mohr, 1980.

———. "Die Rolle der Erfahrung im religiösen Leben nach dem heiligen Bernhard." *Analecta Cisterciana* 46 (1990):307-19.

Lakoff, George, and Mark Johnson. *Metaphors We Live By*. Chicago: University of Chicago Press, 1980.

Landsberg, P. "Les sens spirituels chez S. Augustin." *Dieu vivant* 11 (1948):81-105.

Leclercq, Jean. "Origène au XIIe siècle." *Irenikon* 24 (1951):425-39.

———. *Receuil d'études sur saint Bernard et ses écrits*. Rome: Edizioni di storia et letteratura, 1962-1969.

———. *Bernard of Clairvaux and the Cistercian Spirit*. Translated by Claire Lavoie. Cisterican Studies 16. Kalamazoo, MI: Cistercian Publications, 1976.

———. "L'expérience mystique d'après S Bernard de Clairvaux." *Studia Missionalia* 26 (1977):59-71.

———. *The Love of Learning and the Desire for God: A Study of Monastic Culture*. 3d ed. Translated by Catharine Misrahi. New York: Fordham University Press, 1982.

Leclercq, Jean, ed. *Saint Bernard Théologien.* (= Analecta Sacri Ordinis Cisterciensis 9 (1953), fasc. 3-4.) Rome: Editioned Cistercienses, 1953.

Lerner, Robert. "The Image of Mixed Liquids in Late Medieval Mystical Thought." *Church History* 40 (1971):397-411.

Lewis, Gertrude Jaron, Frank Willaert, and Marie-José Govers, eds. *Bibliographie zur deutschen Frauenmystik des Mittelalters. Mit Anhang zu Beatrijs van Nazareth und Hadewijch.* Berlin: E. Schmidt, 1989.

Longpré, Ephrem. "L'Eucharistie et l'union mystique selon la spiritualité franciscaine." *Revue d'ascétique et de mystique* 25 (1945):306-33.

———. "Bonaventure, Saint." *Dictionaire de spiritualité* 1, col. 1768-843.

Lubac, Henri de. *Exégèse médiévale: Les quatre sens de l'écriture*, vol. 1. 4 vols. Paris: Aubier, 1959.

Maguire, Joanne M. "Nobility and Annihilation in Marguerite Porete's *Mirror of Simple Souls.*" Ph.D. dissertation, University of Chicago, 1996.

McGinn, Bernard. "Ascension and Introversion in the *Itinerarium mentis in Deum.*" In *San Bonaventura 1274-1974.* 5 vols. Grottaferrata and Rome: Collegio S. Bonavantura, 1974, 3:535-52.

———. "Resurrection and Ascension in the Christology of the Early Cistercians." *Citeaux in de Nederlanden* 30 (1979):5-22.

———. *The Foundations of Mysticism: Origins to the Fifth Century.* New York: Crossroad, 1991.

———. *The Growth of Mysticism: Gregory the Great Through the Twelfth Century.* New York: Crossroad, 1994.

———. "Love, Knowledge, and *Union Mystica* in the Western Mystical Tradition." In *Mystical Union in Judaism, Christianity and Islam: An Ecumenical Dialogue*, edited by Moshe Idel and Bernard McGinn. New York: Continuum, 1996:59-86.

———. *The Flowering of Mystcism: Men and Women in the New Mysticism—1200-1350.* New York: Crossroad, 1998.

McIntosh, Mark A. *Mystical Theology: The Integrity of Spirituality and Theology.* Oxford and Malden, MA: Blackwell, 1998.

Milhaven, John Giles. *Hadewijch and Her Sisters.* Albany: SUNY Press, 1993.

Mommaers, Paul, S.J. "Functie van der taal in de mystieke beleving volgens de *Brieven* van Hadewijch," in *Ons Geestelijk Erf* 61 (1987):135-62.

Mommaers, Paul and Norbert de Paepe, eds. *Jan van Ruusbroec: The Sources, Content and Sequels of His Mysticism.* Medievalia Lovaniensia, ser. 1, stud. 12. Leuven: Leuven University Press, 1984.

Mouroux, Jean. "Sur les critères de l'expérience spirituelle d'apres les Sermons sur le Cantique des Cantiques." In *Saint Bernard Théologien: Analecta Cistercana* 9 (1953):251-67.

Murk-Jansen, Saskia. *The Measure of Mystic Thought: A Study of Hadewijch's Mengeldichten*. Göppingen: Kümmerle Verlag, 1991.

———. "The Mystic Theology of the Thirteenth-Century Mystic, Hadewijch, and Its Literary Expression." In *The Medieval Mystical Tradition in England*, 5, edited by Marion Glasscoe. Woodbridge, Sussex and Rochester, NY: D.S. Brewer, 1992:117-27.

———. "The Use of Gender and Gender-related Imagery in Hadewijch." In *Gender and Text in the Later Middle Ages*, edited by Jane Chance. Gainesville, FL: University Press of Florida, 1996:52-68.

———. *Brides in the Desert: The Spirituality of the Beguines*. Maryknoll, NY: Orbis Books, 1998.

Nautin, Pierre. *Origène. Sa vie et son oeuvre*. Paris: 1977.

Newman, Barbara. "Hildegard of Bingen: Visions and Validation." *Church History* 54 (1985):163-75.

———. "*La Mystique Courtoise*: Thirteenth-Century Beguines and the Art of Love." In *From Virile Woman to WomanChrist: Studies in Medieval Religion and Literature*. Philadelphia: University of Pennsylvania Press, 1995:137-67.

Olphe-Galliard, M. "Les sens spirituels dans l'histoire de la spiritualité." In *Nos sens et Dieu*, edited by Claude Baudoin et al. Études Carmélitaines. Desclée de Brouwer, 1954:179-93.

Ong, Walter. *The Presence of the Word: Some Prolegomena for Cultural and Religious History*. Minneapolis: University of Minnesota Press, 1967.

Paulsell, Stephanie. "Writing and Experience of God in the Works of Marguerite d'Oingt." Ph.D. dissertation, University of Chicago, 1993.

Penco, Gregorio "La dottrina dei sensi spirituali in Gregorio Magno." *Benedictiana* 17 (1970):161-201.

Posset, Franz. *Christi dulcedo*: The 'Sweetness of Christ' in Western Christian Spirituality." *Cistercian Studies* 30 (1995):245-65.

Poulain, Auguste. *The Graces of Interior Prayer*. Translated by Leonora L. Yorke Smith. London: Kegan Paul, 1910; reprint, St. Louis, MO: Herder, n.d. Translation of *Les grâces d'oraision*, 6th ed. Paris: 1901.

Rahner, Karl. "Le début d'une doctrine des cinq sens spirituels chez Origène." *Revue d'ascetique et de mystique* 13 (1932):112-45. Published in English, abridged, as "The 'Spiritual Senses' according to Origen" in *Theological Investigations* 16:81-103. New York: Seabury, 1979.

———. "La doctrine des 'sens spirituels' au Moyen Age, en particulier chez saint Bonaventure." *Revue d'ascetique et de mystique* 14 (1933):263-99. Published in English, abridged, as "The Doctrine of the Spiritual Senses in the Middle Ages" in *Theological Investigations* 16:104-34. New York: Seabury, 1979.

Reynaert, Joris. *De Beeldspraak van Hadewijch*. Studien en tekstuitgaven van Ons Geestlijche Erf 21. Tielt and Bussum: Uitgeverij Lannoo, 1981.

———. "Ruusbroec en Hadewijch." *Ons Geestelijke Erf* 55 (1982):193-233.

———. "Hadewijch: Mystic Poetry and Courtly Love." In *Medieval Dutch Literature in its European Context*, edited by Erik Kooper. Cambridge Studies in Medieval Literaure 21. Cambridge, UK: Cambridge University Press, 1994:208-25.

———. "Mystische bibelinterpretation bei Hadewijch." In *Grundfragen christlicher Mystik*, edited by Margot Schmidt and D. Bauer. Mystik in Geschichte und Gegenwart Series, no. 1: Christliche Mystik 5. Stuttgart-Bad Cannstatt: Frommann- Holzboog, 1987.

Reypens, L[eonce], S.J. "Ruusbroec-Studien. Het mystieke 'gherinen.'" *Ons Geestelijke Erf* 12 (1938):15-86.

Rius-Camps, Josep. "Origenes y Marción: carácter preferemente antimarcionita del prefacio y del segundo ciclo del Peri Archon." In *Origeniana*, edited by H. Crouzel et al. Bari: Instituto di Letteratura Christiana Antica, 1975:297-312.

Robertson, Duncan. "The Experience of Reading: Bernard of Clairvaux, Sermons on the Song of Songs, I." *Religion and Literature* 19 (1987):1-20.

Rubin, Miri. *Corpus Christi: The Eucharist in Late Medieval Culture*. Cambridge: Cambridge University Press, 1991.

Ruh, Kurt. "Beginenmystik. Hadewijch, Mechthild von Magdeburg, Marguerite Porete." *Zeitschrift für deutsches Altertum in detsche Literatur* 106 (1977):265-77.

———. "Gottesliebe bei Hadewijch, Mechthild von Magdeburg und Marguerite Porete." In *Romanische Literaturebeziechungen im 19. und 20. Jahrhundert*, edited by Angel San Miguel, R. Schwaderer, and M. Tietz. Tübingen: Gunter Narr Verlag, 1985:243-54.

———. *Geschichte der Abendländischer Mystik*. 4 vols. Vol. 2, *Frauenmystik und Franziskanische Mystik der Frühzeit*. Vienna: Mohr, 1993.

Ruh, Kurt, ed. *Abendländischer Mystik im Mittelalter*. Stuttgart: J.B. Metzlersche, 1986.

Sacks, Sheldon, ed. *On Metaphor*. Chicago: University of Chicago Press, 1978.

Schalij, J.M. "Richard van St. Victor en Hadewijchs 10de Brief." *Tijdschrijft voor Nederlandse Taal- en Letterkunde* 62 (1943):219-28.

Scheepsma, Wybren. "Hadewijch und die *Limburgse Sermonen:* Überlegungen zu Datierung, Identität und Authentizität." In *Deutsche Mystik im abendländischen Zusammenhang: Neu erschlossene Texte, neue methodische Ansätze, neue theoretische Konzepte—Kolloquium Kloster Fischingen 1998,* edited by Walter Haug. Tübingen, 2000:653-82.

Schmidt, Margot. "Rudolphe de Biberach." *Dictionaire de Spiritualité* 13, col. 846-50.

———. "Elemente der Schau bei Mechthild von Magdeburg und Mechtild von Hackeborn: Zu Bedeutung der geistlichen Sinne." In *Frauenmystik im Mittelalter,* edited by P. Dinzelbacher and D. Bauer. Ostfildern: Schwabenverlag, 1985:123-51.

Scholl, Edith. "The Sweetness of the Lord: *Dulcis* and *Suavis.*" *Cistercian Studies Quarterly* 27 (1992):359-66.

Sells, Michael. *Mystical Languages of Unsaying.* Chicago and London: University of Chicago Press, 1994.

Sharf, Robert. "Experience." In *Critical Terms for Religious Studies,* edited by Mark C. Taylor. Chicago and London: University of Chicago Press: 1998:94-116.

Smith, Susan Warrener. "Bernard of Clairvaux and the Nature of the Human Being: The Special Senses." *Cistercian Studies Quarterly* 30 (1995):3-13.

Smith, Jonathan Z. *Drudgery Divine.* Chicago: University of Chicago Press, 1990.

Southern, Richard. *The Making of the Middle Ages.* New Haven: Yale University Press, 1959.

Stock, Brian. "Experience, Praxis, Work and Planning in Bernard of Clairvaux: Observations on the *Sermones in Cantica.*" In *The Cultural Context of Medieval Learning: Proceedings of the First International Colloquium on the Philosophy, Science and Theology of the Middle Ages,* edited by J. E. Murdoch and E. D. Scylla. Boston Studies in the History of Science Series. Dordrecht: Reidel, 1976:219-68.

Straw, Carole. *Gregory the Great: Perfection in Imperfection.* Berkeley and Los Angeles: University of California Press, 1988.

Thomas, Amy. "But What About the Oxen? Body, Soul and Spirit in Origen's Exegesis." Unpublished paper.

Torjesen, Karen Jo. *Hermeneutical Procedure and Theological Method in Origen's Exegesis.* Berlin: de Gruyter, 1986.

Tyler, Stephen. "The Vision Quest in the West, or What the Mind's Eyes Sees." *Journal of Anthropological Research,* 40 (1983):23-40.

Van Mierlo, Jozef. "De 'Minne' in de Strophische Gedichten." *Verslagen en Mededelingen van de Koninklijke Vlaamse Academie voor Taal- en Letterkunde* (1941):687-705.

Vauchez, André. *The Laity in the Middle Ages*. Edited by Daniel E. Bornstien, translated by Margery J Schneider. Notre Dame, IN: University of Notre Dame Press, 1993.

Vekeman, H. "Angelus sane nuntius: Een interpretatie van het Visioenenboek van Hadewijch." *Ons Geestelijk Erf* 50 (1976):225-59.

Verdeyen, Paul. "William of Saint Thierry's Influence on the Flemish Mystics." In *William, Abbot of St Thierry*, edited by L. Mills et al. Kalamazoo, MI: Cistercian Publications, 1987:240-53.

Verwijs, E.-J., and J. Verdam, eds. *Middelnederlandsch woordenboek*. 9 vols. and 2 supplements. The Hague: Nijhoff, 1885-1952.

Von Ivánka, Endre. "L'union à Dieu: La structure de l'âme selon S. Bernard." In *Saint Bernard Théologien* (Analecta Cisterciensia 9 [1953]), edited by Jean Leclercq. Rome: Editiones Cistercienses, 1953:202-8.

Walsh, James. "Guillaume de Saint-Thierry et les sens spirituels." *Revue d'ascetique et de mystique* 35 (1959):27-42.

Webster, Marilyn W. "Mechthild of Magdeburg's Vocabulary of the Senses." Ph.D. dissertation, University of Massachusetts, Amherst, 1996.

Wiethaus, Ulrike. "Learning as Experiencing: Hadewijch's Model of Spiritual Growth." In *Faith Seeking Understanding*, edited by G. Berthold. Manchester, NH: St. Anselm College Press, 1991:89-106.

Willaert, Frank. *De poetica van Hadewijch in de Strofische Gedichten*. Utrecht: DES, 1984.

Willaert, Frank. "Hadewijch und ihr Kreis in den 'Visionen'." In *Abendländische Mystik im Mittelalter: Symposion Kloster Engelberg, 1984*, edited by Kurt Ruh. Series: Germanistische Symposien Berichtsbände 7. Stuttgart: J. B. Metzlersche, 1986:368-87.

Wiseman, James A. "*Minne* in *Die Gheestelike Brulocht*." In Paul Mommaers and Norbert de Paepe, eds., *Jan van Ruusbroec: The Sources, Content and Sequels of His Mysticism*. Leuven: Leuven University Press, 1984:86-99.

Wolfson, Elliot R. *Through a Speculum That Shines: Vision and Imagination in Medieval Jewish Mysticism*. Princeton: Princeton University Press, 1995.

Index

For Product Safety Concerns and Information please contact our EU representative GPSR@taylorandfrancis.com
Taylor & Francis Verlag GmbH, Kaufingerstraße 24, 80331 München, Germany

www.ingramcontent.com/pod-product-compliance
Ingram Content Group UK Ltd.
Pitfield, Milton Keynes, MK11 3LW, UK
UKHW022119190926
475702UK00013B/1541
9780415867023